Cambridge Elements ≡

Elements in the Philosophy of Science
edited by
Robert Northcott
Birkbeck, University of London
Jacob Stegenga
University of Cambridge

SCIENTIFIC KNOWLEDGE AND THE DEEP PAST

History Matters

Adrian Currie
University of Exeter

CAMBRIDGE
UNIVERSITY PRESS

CAMBRIDGE
UNIVERSITY PRESS

University Printing House, Cambridge CB2 8BS, United Kingdom

One Liberty Plaza, 20th Floor, New York, NY 10006, USA

477 Williamstown Road, Port Melbourne, VIC 3207, Australia

314–321, 3rd Floor, Plot 3, Splendor Forum, Jasola District Centre,
New Delhi – 110025, India

79 Anson Road, #06–04/06, Singapore 079906

Cambridge University Press is part of the University of Cambridge.

It furthers the University's mission by disseminating knowledge in the pursuit of education, learning, and research at the highest international levels of excellence.

www.cambridge.org
Information on this title: www.cambridge.org/9781108730556
DOI: 10.1017/9781108582490

First published 2019

A catalogue record for this publication is available from the British Library.

ISBN 978-1-108-73055-6 Paperback
ISSN: 2517-7273 (online)
ISSN: 2517-7265 (print)

Scientific Knowledge and the Deep Past

History Matters

Elements in the Philosophy of Science

DOI: 10.1017/9781108582490
First published online: August 2019

Adrian Currie
University of Exeter

Author for correspondence: Adrian Currie A.Currie@exeter.ac.uk

Abstract: Historical sciences such as paleontology and archaeology have uncovered unimagined, remarkable and mysterious worlds from the deep past. How should we understand the success of these sciences? What is the relationship between knowledge and history? In *Scientific Knowledge & the Deep Past: History Matters,* Adrian Currie examines recent paleontological work on the great changes that occurred during the Cretaceous period – the emergence of flowering plants, the splitting of the mega-continent Gondwana, and the eventual fall of the dinosaurs – to analyse the knowledge of historical scientists, and to reflect upon the nature of history. He argues that distinctively historical processes are 'peculiar': they have the capacity to generate their own highly specific dynamics and rules. This peculiarity, Currie argues, also explains the historian's interest in narratives and stories: the contingency, complexity and peculiarity of the past demands a narrative treatment. Overall, Currie argues that history matters for knowledge.

Keywords: Paleontology, Archaeology, Narrative, Explanation, Evidence

ISBNs: 9781108730556 (PB), 9781108582490 (OC)
ISSNs: 2517-7273 (online), 2517-7265 (print)

Contents

Introduction

Many of science's most revolutionary discoveries concern *deep time*: the past stretching beyond memory and written texts. One discovery was deep time itself. In Europe at least, the realization that the Earth outruns by millennia both biblical history and our own species' existence shook conceptions of humanity's place in the world just as surely as earlier astronomical discoveries (Rudwick, 2014). But the past isn't simply long. Before Homo sapiens, the climate and continents shifted while diverse lineages arose, became extinct and others evolved. We are the result of millions of years of evolution, a heritage that shapes and constrains how we adapt to our still-changing world. Before written records, then, there wasn't simply a past – there was *history*. Scientific understanding of the deep past emerged in the nineteenth century and crucial aspects, how plate tectonics shape geography and climate for instance, have only been accepted in the last half-century (Oreskes, 1999). Considering the extent to which extinction, evolution, plate tectonics and deep time itself form the furniture of our conceptions of the world and our place within it, their recent pedigree is startling.

In what follows, we'll examine the nature and epistemology of this 'deep past'; what we might call 'prehistorical history'. How do historical scientists reach beyond human memory? How does the nature of the past constrain our knowledge? How does history matter for knowing?

Our central question, then, concerns the relationship between history and knowledge. In making inferences, as well as understanding and explaining the world, does history matter? There are at least two ways in which it might. First, is there something special about trying to understand processes or entities located in the past, as opposed to in the present or future? That is, are there systematic claims to be made about the epistemic status of past things in virtue of their being past? Second, does something's history matter to the knowledge we can have of it? The former question concerns whether a target's being in the past makes it special qua object of knowledge. The latter concerns whether a target's history matters for how we might come to know it.

Regarding that first question, I'll draw a negative conclusion: processes and patterns in the past do not represent a fundamentally different kind of epistemic target from those occupying different temporal locations. Regarding the second question, I'll draw a positive conclusion, for two reasons: (1) the past matters for all scientific inference because the provenance of an inference's data always matters; (2) history matters because how a target came to be can make a difference to what knowledge we can have of it. Note an ambiguity: by 'history' do we mean simply being past, or do we mean something richer – having a particular kind of past, or instantiating a particular kind of dynamics?

I'll typically reserve the term 'history' for this latter notion: 'history' doesn't so much pick out a temporal location (the past) as it picks out a set of events, entities or processes for whom distinctive, 'historical' features make a difference to how we might know them. What might these historical features be? Well, that question is in a nutshell what this Element is about.

We'll also ask after the epistemic value of such enquiries. I'll suggest that investigating the deep past involves more than considering particular histories: we delve into the great diversity of forms, structures, trajectories, events, entities and processes that constitute and shape the world, and the conditions enabling them; we bring contemporary conceptions to the past in analysis, data-gathering and interpretation, and that past in turn shapes those conceptions. Thus, history matters for knowledge, and the process of understanding the deep past is rich and invaluable.

Finally, knowing about the past is necessary for understanding much of what we'd like to know. We occupy an often unrepresentative, atypical sliver of time. Our immediately accessible sample is biased, extremely incomplete, inadequate to answer questions at long scales (Marshall, 2017). These questions are Big. How does evolution work? How do planets, solar systems and the universe form? What explains geographical patterns: mountain ranges, valleys? How did our species evolve and radiate across the globe? And these questions matter: How do species become extinct, how do changes in atmospheric composition alter global temperatures and how do changes in global temperatures affect everything else? Answering these questions requires evidence and perspectives that overcome the inherent bias of our little sliver: a long-term view into the deep past. History matters at least because knowledge of it is necessary for answering Big Questions.

This Element is divided into three related parts. In the first, I consider the relationship between evidence, justification and the deep past. In the second, I consider the nature and contingency of history. In the third, I turn to narrative explanation and its role in history. I conclude with a discussion of the value and purpose of historical science itself. I haven't quite written an introduction to the philosophy of the historical sciences. It is instead an extended essay on the relationship between history and knowledge. Given the aims of the Elements series – to be accessible yet substantive – I've aimed for ambition over completeness. Better, I think, to push the boat into deep waters and risk foundering than to stick to cautious shallows.

1 History and Evidence

History and evidence are intertwined: to ask certain questions we need a long-term scale; to uncover the past we require at least some of its remnants to join us in the present. But is there something special about historical evidence: is it particularly difficult, or impoverished or privileged? Some have thought so.

Recently, Derek Turner (2005, 2007) argued that historical evidence is systematically less powerful than experimental evidence; Carol Cleland (2002, 2011, 2013) argued that historical evidence underwrites a distinctive method that is at least equal to more familiar forms of scientific knowledge. Such arguments appeal to some fundamental ontological or epistemological differences between the past on the one hand, and the present and future on the other. For Turner, investigating the past denies us the boon of repeated experimental investigation; for Cleland, investigation of the past grants the boon of bountiful traces.

I'll argue that although in historical contexts evidence's past matters, this is true of all evidence, and so carries with it no special insight about historical knowledge per se. There is nothing distinctive, epistemically speaking, about past objects of knowledge. I'll begin with a methodological discussion, arguing that to understand how history matters for knowledge, we should begin by understanding the practices that generate such knowledge. This motivates examining historical reasoning 'in play': we'll look at recent work on dinosaur development. Based on that case study, I'll then characterize 'trace-based reasoning', presenting and resolving a puzzle concerning it, before considering the relationship between experimental reasoning and trace-based reasoning. I'll conclude that no evidential reasoning escapes history; however, a target's being in the past doesn't in and of itself raise distinctive epistemic challenges.

1.1 The Very Possibility of Historical Knowledge

Where should our project begin? That is, to understand the relationship between knowledge and the deep past, which philosophical approach is appropriate? I think our starting point should be the practices of historical scientists themselves, but let's consider a few options to see why.

Perhaps we should ask after historical knowledge's very possibility: what is necessary for justified knowledge about any past occurrence? Roughly a century ago, Bertrand Russell posed a thought experiment (Russell, 1921). Imagine that the world blinked into existence five minutes ago: a world which is in every way identical to the world as it is, except its past is only 300 seconds long. An identical duplicate, containing all the same memories, fossils, elementary particles and so forth, as our world. Can we tell whether we live in a billions-of-years-old or 300-seconds-old universe? As Russell points out,

There is no logical impossibility in the hypothesis that the world sprang into existence five minutes ago, exactly as it then was, with a population that 'remembered' a wholly unreal past. There is no logically necessary connection between events at different times; therefore nothing that is happening now or will happen in the future can disprove the hypothesis that the world began five minutes ago (Russell, 1921, p. 158).

All present evidence *underdetermines* the hypothesis that the world has a long past and the hypothesis that the world blinked into existence 5 minutes (or 20 minutes, or 3 seconds) ago. 'Underdetermination' is a relationship between at least two hypotheses and a body of evidence: when the evidence is insufficient to decide between those hypotheses, they are underdetermined by it (Wylie, 2019; Godfrey-Smith, 2008; Laudan, 1990). Under Russell's scenario any observations we might make now are the same whether or not the past existed for billions of years or 300 seconds. However, under the sceptical hypothesis, all claims about the past beyond those five minutes appear to come out false, while under the standard hypothesis some at least are true. In one world it is approximately true that (non-avian) dinosaurs died out 65.5 million years ago, in the other, (ignoring philosophical finagling about the nature of truth and reference) it is false.

What are we to make of such hypotheses? We might take our cue from Russell himself: 'I am not here suggesting that the non-existence of the past should be entertained as a serious hypothesis. Like all sceptical hypotheses, it is logically tenable but uninteresting' (Russell, 1921, pp. 159–160).

I don't think Russell is quite right about this.[1] One take-home message is that our knowing about the past depends on features of the past itself: in this extreme case it depends on the past existing (or having had existed). If our knowledge of the past is to be like other empirical knowledge, then it is predicated on there being something that our knowledge is about. Russell is right that, as a serious hypothesis, the five-minute-hypothesis is rather uninteresting. After all, if no possible observation might make a difference to our status as knowers, what are we to do but shrug? But the thought that knowledge of the past depends in part on its nature deserves reflection (which I'll turn to in Section 2).

Russell's sceptical hypothesis is an extreme version of a common philosophical approach. Such approaches begin by generating an epistemic demand: a philosophical bar is set, and knowledge-claims are checked to see if they can make the jump. Some claims, or sets of claims, might be high-jumpers, while others flop (and not in the Fosbury sense). We proceed by asking who makes it

---

[1] Russell's thought experiment is intended only to demonstrate that induction cannot rest on logical, deductive grounds.

over the bar and who fails? Can we answer Russell's sceptic? These are grand, even gallant, starting places – but I think they aren't the only (nor the most productive) opening salvos we might make.

Perhaps we should start by asking whether there are fundamental differences between the past, present and future. If such fundamental differences are to be had, these could constrain or enable different kinds of knowledge or routes to knowledge. Here, we begin by considering the ontology of the past, in contrast to the present and future. Carol Cleland (2002, 2011) argues for something like this. She claims there is a physical asymmetry between the present's relationship to the past and the present's relationship to the future. Causes have multiple effects, and these spread through time. This multiplying aspect, according to Cleland, leads the present to *over*determine the past. That is, the way things are now is more than sufficient to guarantee the way things were, but not so the future, hence the asymmetry. Arthur Danto (1962) says something similar. The fixity of the past, for him, stands in stark contrast with the open future. And in virtue of this, our retrospective understanding of the past is simply of a different nature from our capacity (or lack thereof) to predict the future.

For Cleland, the temporal asymmetry underwrites the method historical scientists adopt. An approach that similarly ties together the relationship between the past and present, and scientific method, is *uniformitarianism*. The term was coined by the nineteenth-century geologist Charles Lyell (1837) and has undergone various – often increasingly complex – incarnations (see Gould, 1965; Camardi, 1999, although I'm basing my discussion on Rudwick's treatments, 1972, 2014). For our purposes, we can divide the idea into two claims: *actualism* and *gradualism*. Actualism claims that our knowledge of the past needs to be grounded in examinations of processes acting today: it is about how to get knowledge. Gradualism claims that processes in the past occurred at roughly the same rate as they occur today. Specifically, change is slow, incremental: great mountains and deep seas are formed by slow local changes; biodiversity is caused by patterns of individual survival, birth and death within particular populations. There are merely methodological or pragmatic versions of uniformitarianism that avoid substantive claims about the past, but such positions have nothing to say about the prerequisites of past knowledge.

Taking actualism and gradualism together, then, we have uniformitarianism. Actualism says that current processes are the key to past processes, and gradualism says that geological features are the result of those processes acting incrementally on grand scales; small changes scaled-up explain big changes: water trickles form mighty gorges.

Understood in a sufficiently weak way, actualism is important. To make inferences about the past, we need *traces*. We need to find remains – footprints, fossils, cave paintings, droppings and so on. Further, to make those inferences we need to understand the processes that shape the past, and examining current processes is one way of achieving this. Although I think taking traces as the basis of our knowledge about the past can be (and typically is) taken far too far (see Currie, 2018a, chapters 6-11), and that there are many ways in which past processes might differ from current processes, no doubt we have to start somewhere. Here, then, is a beginning for our examination of historical knowledge: 'Knowing about the past requires taking the present as having been shaped by its past—that is, to contain some kind of record which we can either decode, or perhaps decode one day'.[2]

Such a thought underwrites the trace-based reasoning we'll meet below.

Uniformitarianism's other half – gradualism – fares worse. Gradualism claims that change is slow and incremental; however, there are plenty of exceptions. Let's glance at a geological and then a biological example. 'Outburst floods' are enormous floods occurring when previously dammed lakes are freed. Such floods shaped North America's distinctive geology, often as a result of ice-age glacial blockades melting. This freed superlative amounts of water, which gouged out massive valleys and were partly responsible for the layers of soft and hard stone necessary for Badlands to form, as well as potentially changing weather patterns (Kehew & Teller, 1994). These are anything but gradual processes.

In biology, gradual speciation has been challenged in two ways. The most obviously gradual model of speciation is *anagenic*: one species gradually shades into another. We can contrast this with *cladogenic* speciation, when speciation occurs by 'splitting' two populations. There is considerable ongoing debate over to what extent speciation follows each process, but I'd be surprised if one model dominates (Plutynski, 2018). On a macroevolutionary scale, we can compare *phyletic* speciation – the thought that speciation occurs with gradual, steady change – with *punctuated equilibrium* (Gould & Eldredge, 1993). On the latter view, over evolutionary time most species are typically in stasis but when change comes, it comes dramatically. Again, there is debate about to what extent stasis or change is the dominant pattern in evolution, and to what extent speciation is gradual or punctuated. In light of these kinds of examples we can't just assume gradual speciation, nor, I think, can we assume gradualism more generally.

---

[2] Derek Turner would point out the explicit textual metaphor here!

Approaching the relationship between history and knowledge by starting with in-principle constraints on the possibility of knowing the past, or considering the relationship between the past and present abstractly, has a set of attendant risks. Such accounts often commit to a particular analysis of knowledge, or to particular metaphysical views, making their story about historical knowledge beholden to the fate of those commitments. I think we should avoid hitching our wagon to a particular analysis of the nature of knowledge or of the relationship between past and present. There is a danger of being fatally disconnected from the phenomena we're trying to understand. The hard-won, transformative historical knowledge that impressed us in the introduction was achieved via human ingenuity and sweat; it wasn't bequeathed from some general fundamental fact about the world, and the phenomenon holds independently of our solving philosophical puzzles about knowledge. The task of actually explaining how historical scientists successfully investigate the often obscure, often weirdly alien recesses of the deep past remains even if we've answered such questions. Under these conditions, striving for philosophical grounding can become a distraction, a red-herring, a roadblock to understanding.

I think a better approach starts with scientific practice: that is, we should examine what scientists actually do. Instead of doing philosophy first – setting a philosophical standard and examining practices to see if they meet it – we start with an examination of scientific work. How do historical scientists reason? What kinds of evidence do they provide? What knowledge-generating processes – fieldwork, experimentation, etc. . . . – do they engage in? After examining science, we see what philosophical systematization and lessons might be drawn from these practices. As such, we'll start by delving into some paleontology in the next Section.

You might complain: if my philosophical analysis begins with a descriptive case study, can I do the explanatory, normative work philosophy demands? If my criteria for good evidence are derived from descriptions of practice, don't I face a methodological dilemma? On one horn, my analysis is restricted to mere description, thus falling short of my normative goals; on the other horn, I fall afoul of the dictum that one mustn't derive an ought from an is. I think this complaint is mistaken: my discussion is neither purely descriptive nor normative and (or so I hope) avoids problematic circularity (Currie, 2015). To see why, I invite you to come along for the ride. In my concluding discussion I'll suggest that our reflections on the nature of the deep past and our knowledge of it are applicable to philosophy, and in a way that enables escape from this dilemma.

1.2 Growing Up Dinosaur

The archaeologist Christopher Hawkes saw historical texts as a crucial evidential crutch supporting old-world archaeology. Even if the target culture did not leave a written tradition, if it is in some way continuous with a culture which did, that continuity can underwrite rich reconstructions of cultural pasts: 'In rural economy, burial rites, technology, sociology or what not, there is always, somewhere or other, a point of reference within the historic order' (Hawkes, 1954, p. 160).

For Hawkes, insofar as we can rely on (and stretch) written records – find a point of reference – we can make inferences from material remains to specific, contingent features of past human societies. For the biology of the deep past, living descendants are the equivalent of historical texts. If an extinct critter has close relatives in the present, then examining those relatives can be an often powerful guide, providing a point of reference within living biota.

This is what makes dinosaurs so challenging.

Dinosaurs' closest living relatives are their progeny, the birds, and their cousins, the crocodilians. These lineages share a common ancestor around 240 million years ago, in the midst of the Triassic period (Green et al., 2014). Although birds and crocodiles provide some guidance for dinosaur palaeontologists, millions of years of evolution opened wide morphological, physiological and behavioural differences between these lineages. Neither living crocodiles nor birds, for instance, include in their ranks multi-ton, terrestrial, herd-living herbivores. If we wish to understand these critters, the horned and frilled ceratopsids, plumed Hadrosaurs, entanked ankylosaurs and earth-shaking sauropods, do we look to birds and crocodiles, or to their mammalian analogues (elephants and hippopotami), or to some combination of both? Are triceratops and friends more like scaled-up, wingless birds, or like reptiles playing at mammalhood? Are they somewhere in between? Are they something else entirely? To see these difficulties in play and how scientists respond, let's consider dinosaur 'ontogenetic development': their patterns of growth.

It is misleading to think that paleontology 'starts' with fieldwork, but our examination of dinosaur development has to begin somewhere. Fieldwork is not simply a matter of finding fossils. Specimen discovery is often challenging, and requires decisions about where to look and what is worth digging up. Fossil extraction is typically destructive. Fossils (particularly of larger animals) often weigh many tons, and don't typically hang out in convenient locations. Of those fossils that are retrieved, decisions must be made about storage and preparation. Preparing fossils is necessary for them to be analysed: only once the biological signal of the fossil has been split from the surrounding rock, can we discern and measure morphology (Wylie, 2015). And fossil preparation is an onerous task.

Post-preparation, decisions must then be made about which prepared fossils are worth analysing, how they should be stored and so forth. Each step, finding the fossil, deciding to dig it up, preparing it and then analysing it, require judgements about how to spend limited resources towards several – often conflicting – goals (Turner, 2016). Building data sets often involves specimens that were dug up and prepared with different practices and different questions in mind. And sometimes these differences make reconciling old data with new difficult (Wylie, 2017). Further, each step introduces a new possible source of bias in the eventual data (Wylie, 2019). The journey from discovery to evidential use is multi-stepped, and each step matters. Often a specimen having been extracted and prepared isn't sufficient for its use as evidence: palaeontologists adopt standards designed to preserve the authenticity and epistemic properties of fossil remains. Leading paleontological journals, for instance, do not accept new species on the basis of privately owned fossils, partly for ethical, partly for epistemic reasons (Havstad, 2019). Much of these complexities are encapsulated in Jack Horner and Mark Goodwin's investigation of how *Triceratops* grew.

Triceratops are instantly recognizable, three-horned ceratopsids from the North American Cretaceous. Although *Triceratops* specimens are relatively common (for dinosaurs), back in 2006 only four non-adults were described in published literature. This was partly because 'smaller *Triceratops* skulls and cranial elements were apparently overlooked, deemed highly incomplete or undesirable to collect' (Horner & Goodwin, 2006, p. 2757). Non-adult skulls are often more brittle, so less likely to survive the fossilization process, and those which do survive are typically incomplete. So if you're looking for the 'best' – most complete – skulls, then you'll focus on adults. The aims and standards of collecting affect what is collected (Wylie, 2017). Further, many of the *Triceratops* prepared in the past were prepared towards ends other than understanding their growth: ' … many previously collected *Triceratops* skulls in museum collections have undergone extensive restoration, are composites or lack contextual field documentation, making their use unreliable' (p. 2757).

In 1997, crews from the Museum of the Rockies and the University of California Berkeley began working the Hell Creek Formation in eastern Montana, aiming for a collection suitable for studying *Triceratops* development. The specimens were prepared under Horner and Goodwin's supervision, providing 10 full and 28 partial skulls.

Sorting the skulls by size and other signals of age, Horner and Goodwin hypothesized a four-stage sequence taking us from infant, juvenile, sub-adult to adult *Triceratops* (see Figure 1).

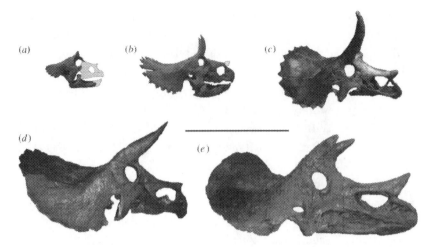

Figure 1 Horner and Goodwin's triceratops growth sequence. (Horner &
Goodwin 2006, 2760) © Royal Society.

Even to uneducated eyes, the changes over a *Triceratops'* lifetime are striking. Most obvious are changes in the 'frill' (the bony crest jutting out the back of
the skull) and in the 'post-orbital' horns (the two horns above the eyes). The frill
begins rather unambitiously, jutting straight out the back in infants (Figure 1a),
to increasingly splayed and dramatic across juveniles and sub-adults (Figure 1b,
c). More strikingly, the post-orbital horns jut upwards with a slight backwards
lean in early life stages (Figure 1a–c), but change dramatically in adulthood
(Figure 1d, e), bending forwards over the eye.

Why does understanding *Triceratops* growth matter? What motivated nine years
of collection, preparation, analysis and publishing? The reconstruction underwrites
further speculation on Horner and Goodwin's part. They suggest that changes in
post-orbital horns are 'probably visual cues of immaturity' (p. 2761). A quick
glance at horn position would be a useful way for *Triceratops* to gauge age. More
dramatically, in 2010, Horner, this time with John Scannella, combined this
sequence with another *ceratopsid, Torosaurus*, to argue that the two were not
different genera after all, but that *Torosaurus* represented the final adult stage
of *Triceratops*. So, the proposed growth sequence provided a basis for further
speculation and led to previous interpretations being re-evaluated.

Further, *Triceratops* skulls inform us about more than ontogeny. Scannella
et al. (2014) used the same skulls to argue that *Triceratops* species evolved by
gradual transformation. Triceratops in the Hell Creek formation have two
morphotypes: *T. horridus* and *T. prorsus*. Instead of arranging skulls by life-
stage, Scannella and company arranged them by stratigraphic sequence (that is,
organized by the layer of rock in which they were discovered). Because strata

are laid down sequentially over time, a stratigraphic sequence can (with some care) be read as a temporal sequence. The pattern reinforced *T. horridus* and *T. prorsus* being separate species, but more interestingly, the scientists also argued that the sequence reveals anagenesis (recall: speciation via gradual transformation of a whole lineage). A cladogenic speciation would express itself in the record differently from an anagenic one. Across the sample there was no stratigraphic overlap and there were apparently transitional forms: highly suggestive of anagenesis.

For Scannella et al., the discovery of anagenetic patterns of speciation in *Triceratops* is significant because it helps set expectations about evolutionary patterns in that lineage. Assuming cladogenic change would lead to a 'problematic inflation of dinosaur diversity' (2014, p. 5). We'll expand on dinosaur biodiversity below, but, first, let's consider another case.

Ceratopsids are not the only late-Cretaceous critters to have undergone ontogenetic study. Another group, also very diverse in the Cretaceous, and also sporting distinctive skulls, are the *Pachycephalosaurids*, a group of hadrosaurs. *Pachycephalosaurs* are famous for their bony, thickened skulls – often represented as head-butting-based male-male competition over mates (à la big horn sheep, although some argue display is just as plausible, Peterson & Vittore, 2012). *Pachycephalosaurid* species are identified by their skulls, which are typically divided into two types. Although all are thick, in one group the skull is shaped to form a dome, in the other the skull is flat. How are these two types related? On one view, they are divided into two clades: *Homalocephale* (the flat-heads) and *Pachycephalosauidae* (the classic dome-heads). On another view, the dome-heads cluster together but the flat-heads do not. Recently, a third option has emerged: once again, an ontogenetic difference has been mistaken for a phylogenetic difference. Following past suggestions (Goodwin et al., 1998; Sullivan, 2003), Schott et al. (2011) analyse the skulls of two well-sampled lineages: the dome-headed *Stegocera validum* and the flat-headed *Ornatotholus browni*. They 'test the hypothesis that *Stegoceras validum* developed ontogenetically from a flat-headed morphology to a domed morphology using multiple independent lines of evidence on a large, stratigraphically constrained sample of specimens . . . from the Belly River Group of Alberta' (p. 2). Let us examine the lines of evidence, and the specimens.

The lines of evidence involve, first, *histology*. Microscopic tissue structure sometimes fossilizes. Examining such structures can provide insight into growth patterns and help reconstruct ontogenetic sequences. For instance, highly vascularized (that is, complex, involving multiple cell-types) cells are often associated with fast – and perhaps later – growth. A second line of evidence exploits 'diagnostic ornamentation': highly conserved features taken to be excellent initial evidence of taxic affiliation. Third, evidence from 'allopatric analyses' of skull shape. That is, analyses that take various features of the skull and map them onto ontogenetic sequences constrained by expectations of growth patterns. Generally speaking, by reconstructing the growth sequence of the adult, Schott et al. are able to show that various traits which dome heads have 'grown out of' are still present in the flat-headed taxa. Fourth, and finally, the size-and-age ranges of the specimens is consistent with an ontogenetic sequence rather than a phylogenetic one. That is, the flat-headed specimens are smaller, and the two overlap in strata.

Schott et al.'s analyses are possible because of their careful assembly of a set of 40 skull specimens, collected from the Dinosaur Park formation in Alberta. Although there are potential samples from New Mexico and Montana sites (where Horner's ceratopsids hailed from) these were left out to 'minimize the chances of sampling multiple taxa' (p. 5). By restricting samples to a relatively constrained location and temporal period, they lessen the chance of noise from the inclusion of yet more species. As in the ceratopsid case, the investigations turn on careful fieldwork, care taken in collections and preparation, and the materials then being stored (often for many years) before finally being deployed as evidence.

Reassessments of dinosaur ontogeny and their taxonomic affiliation relate to more general patterns of biodiversity in the Cretaceous. The Cretaceous period began with a minor extinction event approximately 145.5 million years ago, and ends with the K-Pg extinction event 65.5 million years ago. In addition to the mass extinction closing act, another notable plot point is the *mid-Cretaceous Terrestrial Revolution* (Lloyd et al., 2008; Meredith et al., 2011). Beginning roughly 125 million years ago, land-based flora and fauna profoundly changed. Ferns and gymnosperms (conifers, cycads and their ilk) were replaced by the angiosperms: the flowering plants that now dominate Earth's plant-life. By the late-Cretaceous, angiosperms dominated 80 per cent of terrestrial plant biota (Benton, 2010). Insects and other pollinators quickly radiated into new angiosperm-based niches. Meanwhile, there's evidence that early birds and mammals, as well as the already well-established squamate lizards and crocodilians, underwent major radiations (see Dilcher, 2000 for plants; Grimaldi, 1999 for insects; Hedges et al., 1996 for birds and mammals). This all happened during a major geographical mix-up: the breaking apart of the super-continent Gondwana. In many ways, it was during the mid-Cretaceous that our modern world was formed, not in the crucible of the dinosaur extinction 60 million years later.

It may be that dinosaurs took part in the mid-Cretaceous terrestrial revolution, particularly the herbivores. The Sauropod-dominated Jurassic landscape was replaced by ceratopsids, ankylosaur and hadrosaur. However, it has been argued that apparent biodiversity increases in dinosaur are a result of sampling bias (Lloyd et al., 2008). A good understanding of dinosaur biodiversity is critical for telling to what extent they followed the trends of the mid-Cretaceous. It is also critical for resolving a puzzle about Cretaceous dinosaur mega-fauna: they appear to be too biodiverse, given our expectations from other assemblages of large animals (Sullivan, 2006). Based on suggestions from studies of *Pachycephalosaur* and ceratopsids, a solution emerges: perhaps the biodiversity of late-Cretaceous dinosaurs has been inflated. First, as Scannella et al. imply, assumptions of cladogenic speciation could lead transitional forms to be classified as separate species. Second, the unexpected morphological changes dinosaurs underwent during growth could lead to ontogenetic differences being mistaken for phylogenetic differences.

Finally, these re-evaluations of dinosaur biodiversity matter for understanding their extinction. Explaining the role (if it played a role) of the massive impact at the end of the Cretaceous depends in part on whether biota at the time were thriving, in decline, or in some way especially vulnerable. Although it is increasingly thought that biodiversity was relatively stable up until the K-Pg event (Lloyd et al., 2008), there are now suggestions that the community structure of the late-Cretaceous dinosaurs changed in ways that decreased their resilience (Mitchell, Roopnarine & Angielczyk 2012). All of this work depends on sufficiently accurate phylogenetic information, and this requires distinguishing between ontogeny and differing evolutionary history.

These narratives concerning the relationship between the mid-Cretaceous Revolution, the new angiosperm-insect alliance, and the eventual dinosaur extinction, will play a prominent role in Sections 2 and 3. For the remainder of this Section, we'll draw on reconstructions of dinosaur ontogeny to think about historical inferences.

1.3 Trace-based Reasoning

The questions driving this Element are, first, whether we can say anything systematic about a target's epistemic status based on it being in the past; second, whether a target's history makes a difference to what we might know of it. In Section 1.1, we briefly discussed actualism: our knowledge of the deep past is rooted in understanding how the past has shaped the present. If the present is our key to the past, then getting to the past requires intimate understanding of the present. In Section 1.2, examining late-Cretaceous dinosaur development, we saw actualism enacted in concern for the careful treatment of fossils from fieldwork, to preparation, to scientific analysis. In this Section I will characterize 'trace-based reasoning' more carefully, then pose and resolve something of a puzzle: this kind of reasoning seems far too powerful, it is 'unreasonably effective'.

A 'trace' is some downstream remnant of a past event. A rock bears some similarity to biological morphology, and we explain those similarities as resulting from the rock being a causal descendent of a once-living critter. We infer from a *Pachycephalosaur*-skull-shaped rock to past *Pachycephalosaurs*. This capacity relies on knowing how a carbon-based skull could be transformed into other minerals. Generally speaking, to infer from a trace to the past we need some theory that explains how that past event could have led to the current state. Following Lewis Binford, such theory is typically called 'middle-range', or 'midrange' theory (Binford, 1977; Jeffares, 2008; Kosso, 2001). Trace-based reasoning, then, involves inferring to the past on the basis of explaining things we observe today in terms of their being the result of those past events and processes, on the basis of midrange theory (Currie & Killin, 2019).

Successful trace-based reasoners keep multiple plates spinning simultaneously. The history of the trace as a trace – its extraction, preparation and so forth – matters for what it can evidence, and the stability of midrange theory connecting the trace to the past is critical. In using sets of traces, individual similarities and differences matter. Single trace-sets are deployed and re-deployed for different epistemic uses across time, such as when the same set of *Triceratops* skulls are used to represent an ontogenetic sequence and a stratigraphic sequence. This all requires intimate knowledge of the specimens themselves, as well as rich theoretical understanding of how they could form.

Like any induction, trace-based reasoning is vulnerable. Horner and his collaborators' *Triceratops* skulls only included four infants. Given their fragility, it could have been that none fossilized, or were discovered, or survived to be discovered. Further, given the small size of the sample, could those skulls be atypical? And given that the skulls are often incomplete, how confident should we be in their reconstruction? Evidence decays over time. Much is not preserved at all, and specimen extraction is often destructive: removing a fossil from the ground, and preparing it, almost inevitably involves information-loss. Further – as we've discussed – trace-based reasoning relies on understanding how the present is shaped by the past. But in many ways the past isn't like the present. Recall: the most closely related living animals to *Triceratops* are birds and crocodiles, and the largest mammals are likely to differ in many ways from dinosaurs. So, what analogies should we use to model dinosaur development? Crocodilian, avian, or mammalian? These failures of uniformitarianism make justifying theories of dinosaur development tricky.

We have, then, two sources of underdetermination:

(1) Trace decay,
(2) Failures of uniformitarianism.

Related worries concern interpretation. With little constraint on how we interpret traces, 'Xeroxing' could occur. Our interpretation of data is shaped by preconceptions. Xeroxing happens when those preconceptions lead us to interpret sets of traces similarly, creating the impression that those preconceptions are well founded – after all, they have so many instances (Bell, 2015). In Xeroxing, our preconceptions shape data, not vice versa. If data are biased, dispersed and messy, we might expect our interpretive photocopier to reproduce traces in our preconception's image, thus further reinforcing said preconceptions in an epistemically disastrous cycle. Further, our interpretations are likely to be sensitive to a range of idiosyncratic factors: the training we've undergone, the research questions in mind, and perhaps even more personal features, as Joan Gero put it ' ... our assessments of what is significant, what to record, our assigning of data to specific categories, our willingness or ability to make distinctions on any given day, these all vary tremendously from day to day, from mood to mood, from one knowledge state to another, even by a single investigator' (2007, p. 318).

So, there are many reasons to worry about the stability and licence of trace-based reasoning. When data sets are so impoverished, and so apparently hostage to the whim of our theoretical ideas and biases, what assurances can we have about past knowledge?

Nonetheless, some philosophers are impressed by trace-based reasoning: by how often Xeroxing is avoided, underdetermination is overcome, and interpretation is well-grounded, often on the basis of very minimal traces. Horner and Goodwin's analysis of *Triceratops* skulls included only four incomplete infant skulls, preciously few! There are more extreme examples: only a few scraps of bone (initially a single finger bone) revealed an entirely new species of hominids living contemporaneously with (and likely interbreeding with) *Homo sapiens* and Neanderthal: the Devisonians (Krause et al., 2010). I've elsewhere discussed how a single molar was sufficient to identify a new species of long-extinct platypus (Currie, 2018a). How could so few fossils make such a difference? Traces appear to be *unreasonably effective*: given the challenges from underdetermination and interpretation, they shouldn't be so powerful. As Bob Chapman and Alison Wylie put it, ' ... how stubbornly recalcitrant these data can be, no matter how entrenched their assumed meaning ... ' (Chapman & Wylie, 2016, p. 5).

What do I mean by 'unreasonable effectiveness'? The turn of phrase is owed to the physicist Eugene Wigner who, in a 1959 lecture, remarked on the surprising capacity of mathematics to capture the physical world:

> [I]t is important to point out that the mathematical formulation of the physicist's often crude experience leads in an uncanny number of cases to an amazingly accurate description of a large class of phenomena . . . The miracle of the appropriateness of the language of mathematics for the formulation of the laws of physics is a wonderful gift which we neither understand nor deserve (Wigner, 1960, pp. 8–14).

If mathematics is a purely formal, made-up language, why should it so successfully capture the world 'out there'? Something's unreasonable effectiveness suggests that we have either misconceived it or its success. For instance, some argue from mathematics' unreasonable effectiveness to forms of mathematical realism, while others have argued that mathematical power in the natural sciences says more about our own psychology than math or nature (Colyvan, 2015). If traces are unreasonably effective, then either we have made a mistake analysing the nature of trace-based reasoning, or got carried away in characterizing their success.

Alison Wylie (particularly in her collaboration with Chapman) understands the power of traces in terms of the flexibility and 'non-fundamental' nature of archaeological reasoning (Wylie, 1999, 2011; Chapman & Wylie, 2016). The idea that scientific theories are organized into hierarchical systems of dependence is baked into many conceptions of science: a bedrock of theory forms the stable foundation of our knowledge. On this approach we model scientific testing as involving, say, deductive relationships between theories and observations, where the theories are hierarchically organized (Oppenheim & Putnam, 1958). However, historical scientists adopt varying strategies to accommodate, mitigate, and overcome the various kinds of epistemic worries and issues listed above, and these strategies are highly tailored to the specifics of the task, typically involving highly local, often tacit knowledge. Consider this discussion of fieldwork from Leakey and Lewin:

> A fossil hunter needs sharp eyes and a keen search image, a mental template that subconsciously evaluates everything he sees in his search for telltale clues. A kind of mental radar works even if he isn't concentrating hard. A fossil mollusk expert has a mollusk search image. A fossil antelope expert has an antelope search image. . . . Yet even when one has a good internal radar, the search is incredibly more difficult than it sounds. Not only are fossils often the same color as the rocks among which they are found, so they blend in with the background; they are also usually broken into odd-shaped fragments. . . . In our business, we don't expect to find a whole skull lying on

the surface staring up at us. The typical find is a small piece of petrified bone. The fossil hunter's search therefore has to have an infinite number of dimensions, matching every conceivable angle of every shape of fragment of every bone on the human body (Leakey & Lewin, 1992, p. 26).

Fossilhounds use bountiful theory-mediated expertise. They look for surprising, new, or unexpected fossils, often employing well-developed theories of best practice and techniques for searching and extracting in the field. Similar can be said for fossil preparation, as Caitlin Wylie puts it: ' ... preparators present their work and their role as skilful, individualized, and irreplaceable ... each preparator's artistic skill, aesthetic judgement, and creative problem-solving about prepared specimens' appearance [makes a difference to the end product] and thus scientific interpretation' (2015, p. 52).

These skilful, partially autonomous knowledge-producing activities might form chains of inference, but their links are not purely logical, hierarchical relationships. Further, even with good specimens, palaeontologists still need to discover patterns across those specimens and try to explain and understand the data (Alisa Bokulich, 2018, for instance, discusses the nuanced modelling involved in inferring from the fossil record to biodiversity's deep past).

Trace-based reasoning – inferring from remains to their past causes via midrange theory – is then an important part of understanding historical evidence, but is nonetheless misleading. It makes historical inference appear simple. It obscures the complexity of fieldwork, preparation, storage and analysis on the one hand, and de-emphasizes other routes to uncovering the past on the other hand (I discuss this latter point in great detail elsewhere, see Currie, 2018a, chapters 6-10, also Turner, 2013). Chapman and Wylie deny that archaeology has foundations in the sense of a 'theoretical bedrock', and I think this holds for historical science more generally. This lack of foundation enables historical scientists to exploit different kinds of relationships between their evidence. For Wylie, historical science is neither generally unified nor disunified, but is instead a patchwork of local integration and independence. And it is these local independencies and interdependencies that explain its success (Wylie, 1999).

When I say historical knowledge is *local,* then, I have three things in mind. First, the scope of the knowledge is often highly specific. Although historical scientists make use of quite general ideas – how stratigraphic layers form, the relationship between histological patterns and ontogenetic growth – the application of these ideas requires close attention to the specificities of the traces, the context of their discovery and preparation, and so forth. General tools are not simply applied to the case at hand, but fine-tuned and adapted to that local context. Second, the knowledge is often tacit, skilled know-how. As I'll discuss below, the kind of knowledge we saw in fieldwork and fossil preparation also looms large for other aspects of science like theorizing and reasoning as well. Third, the dependencies between knowledge – the chains of inference and justification – are not hierarchically organized, but are a mosaic of interdependencies. It might be tempting to think of more general ideas like stratigraphy as forming the 'base' of historical knowledge. However, as the knowledge's applicability relies on highly detailed understanding of the specific context, the relation is better understood as one of interdependence.

The power of traces to upset deep-seated preconceptions, then, is partly explained by our expectations not being so deep-seated after all. There is no hierarchy of theories determining interpretation, knowledge is local. My notion of a 'methodological omnivore' (Currie, 2015, 2018) attempts to capture a similar idea (see also Bonnin, 2019). This focus on local patterns of justification and reasoning should lead us to change our perspective vis-à-vis the effectiveness of traces. Traces are not more effective than they ought to be, but their epistemic power is best understood at a more fine-grained level than philosophers are used to. It may be surprising that a few scraps of Denisovan bone could be sufficient to establish a whole new human species, but focusing on those scraps obscures the rich background knowledge that goes into supporting and developing the trace-based inference. It is against the backdrop of hardwon knowledge concerning both how to access, treat and analyse genomic data (including the long road to workable ancient DNA analysis, Jones, 2019), and our knowledge of both *Homo sapiens* and Neanderthal genomic sequences, that the traces are used so effectively. As Wylie puts it, ' … on the model of evidential reasoning I have outlined here, much of the action is off-stage. It is at least as crucial to establish the security and relevance of a robust body of background knowledge … as it is to work in the foreground, recovering and recording the material record that survives of an archaeological subject' (Wylie, 2011, p. 339).

The unreasonable effectiveness of traces is an illusion caused by an over-abstract approach to thinking about evidence: simple models of trace-based reasoning. The layers of interpretation taking us from fieldwork, to fossil preparation, to analysis, to publication, is where the epistemic action lies. Traces are not unreasonably effective. They are exactly as effective as our knowledge allows them to be. This doesn't mean that we shouldn't be (as I am) surprised and delighted by the creativity and ingenuity with which historical scientists uncover the past, but rather that we shouldn't think there is any great mystery concerning their success.

1.4 Experiments and History

A common way of understanding historical science is via comparison with experimental science: intervening on isolated systems to establish causal relations and test hypotheses. In this Section I'll compare trace-based reasoning to experimentation, arguing that history (in the sense of having a past) matters as much for the latter as for the former.

Philosophers have said a lot about experiments (certainly more than about trace-based reasoning!). Construing experiments as narrowly in the business of testing theories is a mistake. Experiments form empirical traditions that are often autonomous from scientific theories – they have a 'life of their own' (Hacking, 1983). Experiments are often exploratory, attempting to discover new phenomena, characterize particular systems and generate surprise (Franklin-Hall, 2005; Parke, 2014; Currie, 2018b; Morgan, 2005). Some philosophers want to equate experimentation with models and simulations (Maki, 2005), whereas others keep them apart (Currie & Levy, forthcoming). Some think experimentation is a privileged route to knowledge, whereas others don't (Parke, 2014). Regarding historical science, Carol Cleland argues that experiments and trace-based reasoning are two different, and equally legitimate, ways of generating knowledge. Derek Turner sees experimental knowledge as privileged: after all, the experimentalist may generate replicated, carefully controlled tests on their subjects; they can 'make their own luck', whereas the historical scientist's epistemic fate is beholden to the survival of traces. Ben Jeffares (2008) points out that establishing midrange theory often involves experimentation, thus blurring the lines between experimental and historical science. I've argued that although historical scientists often can't perform experiments, they have ways of mitigating this, moving beyond trace evidence in various ways (Currie, 2018a).

So, there's plenty of philosophical wrangling to be had concerning what experiments are, their purpose and how they compare to historical reconstruction. Here, I'll argue that just as for traces being evidence requires knowledge of the processes which form them, so too do the results of experiments require mediating knowledge concerning the past of the experimental data in question. I'll make three related claims. First, arguments that experimentalists (as opposed to trace-based-reasoners) can generally 'control their own fates' lean on some unlicensed idealization, ignoring the constraints on experimentation as practiced. Second, it is not a subject's being in the past that makes it inaccessible to experiment per se. Third, all evidence – including experimental evidence – is historical in the minimal sense that all inductive inferences require knowledge of the relevant data's past.

These three points lead to the conclusion that whatever the limits of trace-based reasoning and experimentation might be, a target's being in the past does not make a unique systematic epistemic difference. Let's start by characterizing experiments, focusing for simplicity on those aiming to test hypotheses.

The systems experimental scientists explore, from high-energy physics to developmental genetics to psychology, employ highly artificial systems bearing only a passing resemblance to the whirling dervish that is nature. Presumably, then, not just any concrete intervention counts as an experiment (or a successful one), but those where the concrete set up – the experimental system – bears the right kind of relationship to the relevant hypotheses. One view is that the subject of the experiment and the subject of the hypothesis should be the 'same kind of thing'. Some philosophers have emphasized that experimental scientists, in performing their experiments, bring part of the world into the lab: 'experiments are versions of the real world captured within an artificial laboratory environment' (Morgan, 2005, p. 317; Guala, 2002). This matters because if the experiment is to test a hypothesis, where that hypothesis is a generalization about the natural world, then the object of the experiment should fall within the scope of that generalization (Arnon Levy and myself attempt a nuanced view on this, Currie & Levy, forthcoming).

Carol Cleland's notion of experimentation is tied to testing general regularities. An experiment is a concrete set up of test conditions relevant to a theory or hypothesis. An experimental result will either accord with the hypothesis, and thus be confirmatory, or fail to – thus falsifying the hypothesis. However, the results might only seem to confirm or disconfirm, when in reality some mistake in the experiment's design, some hidden variable, or a fluky result, is to blame. So, experimenters repeat their procedures, varying conditions to control for these confounders – false positives for apparently confirmatory evidence and false negatives for apparently falsifying evidence. The experimenter, then (1) is testing some general theory via the results of a controlled physical intervention, and (2) performs repeated, varied interventions to ensure a match between the result and the theory the experiment is intended to test (Cleland, 2002). Based on the discussion above, we might add: (3) the success of the experiment turns in part on the right kind of relationship holding between the experimental system and the natural system the hypothesis concerns.

We have a potential reason to draw apart experiments and history. Perhaps because objects in the past cannot be the subject of replicable experiments, there is a gap between the experimental and the historical. As Cleland says, ' . . . without the ability to manipulate suspect conditions, one is at the mercy of what nature just happens to leave in her wake; sometimes she is generous and sometimes she is stingy, but the bottom line is that you can't fool with her' (Cleland, 2002, p. 485).

I think Cleland is right that many investigators are, as it were, at nature's mercy – unable to use experiments to 'make their own luck'. However, this situation does not result from the target being in the past. Sciences that concern currently existing systems are also sometimes denied the boon of experimentation.

Many sciences study systems that operate at scales which make experimentation difficult, of lesser use or practically impossible. Economics, astrophysics and climate science are some examples. Ecology is another. Ecologists are in the business of understanding the dynamics of species assemblages: predator and prey, mutualisms, producers and consumers, how energy flows through ecosystems. Although ecologists can perform lab-based experiments on 'microcosms' of ecosystems (and, as with historical science, there is plenty of lab work involved, Currie, forthcoming; Odenbaugh, 2006) and can perform limited field experiments (Millstein, forthcoming), these activities differ from repeatedly intervening on an experimental system to test a general theory pertaining to that system. This is because of scale and complexity: ecosystems are simply 'too large' for us to meaningfully control, reproduce and (as it were) bring into the lab.

You might complain that experiments are beyond ecology *in practice* – because of mundane reasons like a lack of engineering power and know-how, or a lack of time and resources – while historical scientists cannot experiment *in principle* because their targets are in the past. I'm not convinced. First, these mundane 'in practice' constraints matter critically to knowledge production. Lab-based experimentation is often expensive, requires diversely skilled people and can be frustratingly difficult. Moreover, experimenters rely on a community recognizing and enforcing best practices and other norms, as well as institutions enabling dissemination and funding. Repeatability is not so immediately or simply achievable, then. Second, I suspect there is a double-standard regarding how we conceive experimental and historical cases. If in considering ecology we are allowed to ignore 'in practice' constraints, then we ought to do the same for historical science. Removing such constraints, it seems to me, would enable historical scientists to perform many experiments. If there are relevant examples of targets still present, or manufacturable (and remember, we're talking 'in principle' here!), then it seems we could expect experimental reach. Where we cannot experiment on non-avian dinosaurs, this isn't because particular subjects are in the past, but because the whole *category* is in the past. As we've seen, a major challenge for dinosaur paleontology, not faced by mammal or bird paleontology, is the lack of any tokens of that type being around now. It is the capacity to be replicated that matters here, and things in the past can often be replicated. Even if it is right that targets in the past – where the whole category is in the past – are unable to be replicated, in practice this is simply one way that experimental access is limited, along with issues of scale, uniqueness and the distribution of tokens. And further, this assumes we are unable to manufacture targets.

And so, I suspect there is a distinction to be had – rough-hewn and idealized, to be sure – between experimental and non-experimental science. Experimental sciences are able to repeatedly manipulate the objects about which their hypotheses refer and this brings a suite of epistemic benefits and challenges. Other sciences have to rely on data-gathering, proxy investigations and other approaches. But this distinction does not track history and non-history. Let's now turn to a similarity between experiment and history.

Wylie's insight about historical inference is that to understand the power of traces, we mustn't abstract too far from the local processes and practices warranting their evidential use. The stability of historical inference is granted not by some fundamental grounding, nor by purely logical argument, but by the complex, often messy and diverse practices of generating, curating and deploying traces as evidence. Presumably, experiments should be treated with the same courtesy, as Sabina Leonelli has recently (Leonelli, 2016, forthcoming).

Most philosophical analysis of evidence is *synchronic*, that is, we consider evidential reasoning while ignoring temporal aspects: experiments generate evidence that test hypotheses, and we employ an atemporal model to understand their relationship. Such models often encourage philosophers to consider the hypothesis and evidence abstracted from the conditions of their generation. But Leonelli urges that understanding evidential reasoning and support requires a *diachronic* perspective. Experiments generate a set of results, which are recorded, transformed, stored, data-based, analysed and occasionally deployed as evidence. And that temporal process (that recording, transforming, storing, data-basing, analysing) matters for the warrant of evidential claims formed on the basis of those data. Provenance always matters.

Leonelli distinguishes between *data* and *evidence* (Leonelli, 2016). 'Data' are the result of some process of generation: an experiment is run, a fossil prepared, etc. Data go on journeys: they are curated, transformed, de-contextualized from the instance of their creation to be placed in tables, recorded in data bases and so forth. Sometimes, data are deployed as evidence – playing a role in supporting some claim. Data are, in part, potential evidence. As such, their provenance, their past, matters critically. The importance of provenance is unmissable when considering trace-based reasoning. Traces are the remains of long-ago processes, entities and events. They have been transformed over time: by geology most obviously, but also in the often destructive processes of fieldwork, preparation and storage necessary for those traces to be deployed. In considering traces, the trace's past – its status as data – is immediately called to mind.

Provenance matters critically for experiments too. Experiments generate data – recordings, results – which are then deployed. Sometimes the temporal distance between generation and deployment is long indeed. And sometimes the same data might be deployed in many different ways. This is most obvious in 'big data' science, where large amounts of information are stored and categorized in data bases, to be deployed via statistical analysis to a plethora of cases (Leonelli, 2016). However, it is also a more-or-less ubiquitous feature of successful science. Studies themselves will be re-used as citations, in meta-analyses and for journalistic reporting. In each of these cases where the data comes from, its provenance, partly warrants its use as evidence. As Leonelli puts it, 'Data are defined by their temporal characteristics as much as by their spatial and morphological ones, and underestimating the challenges and time-scales involved in data processing can disrupt inferential reasoning and invalidate the use of data as evidence'.

Here's the lesson we should draw from Leonelli: *experimental results (data) are traces of past experiments*. When we deploy experimental data to make claims about phenomena, we inevitably make a trace-based inference to the stability, validity (and so on) of the evidence's source and its subsequent journey. In this weak sense, history always matters: distinguishing between the experimental and the historical by claiming that experiments' repeatability allows them to escape history fails. You might insist that nonetheless repeatability grants experimental scientists advantages; however, again, the repeatability of experiments might be over-emphasized by adopting a too-ideal perspective. As I've mentioned, experiments are often costly, hard to run, require significant expertise and so on. As Leonelli puts it,

> experimental results are difficult to control – not only at the point at which they are produced, but most significantly at the point of dissemination, storage, and re-use. Data can disappear or become unusable very quickly if not properly curated: it only takes a destroyed hard disk, a misleading annotation or a postdoc changing jobs. Worries about differential survival of evidence and informational destruction are thus arguably as alive with contemporary data collection in the life sciences as they are for historical sciences and observational data therein.

Even if experimental data may be repeatedly generated in principle, in practice they often are not. And, by contrast, some evidence of the deep past is truly bountiful (consider the Big Bang's background radiation). It seems to me that the epistemic variability within experimental science, and within historical science, is greater than the variability between experimental and historical science. As such, it is unclear whether comparing them in this way is particularly enlightening. Doing so requires abstracting away from local contexts, which I've argued is necessary for understanding the warrant of that reasoning in the first place. You might further object that historical and experimental data come apart insofar as in the former we care about traces in terms of their natural formation (how bone becomes fossil, for instance), whereas in experimentation we care about the specimen's history post-experiment. However, employing experimental evidence requires intimate knowledge of the context in which the data were generated, just as employing traces needs knowledge of the natural conditions that generated them. Other than issues of repeatability (which we've already covered), I don't see how appeal to 'natural' versus 'artificial' generation draws apart the role of the past in historical but not experimental reasoning.

We have, then, a sense in which history matters for evidential reasoning in science. When data are deployed to form an evidential claim, where the data came from, how it was produced for instance, matters. We missed this because of an overly ideal approach to thinking about evidence. Evidential reasoning is historical reasoning. However, the sense of 'historical reasoning' here is shallow: it simply means *having a past*. Historians and historical scientists often mean something richer than this: the things they seek to understand are somehow *especially* historical. Grappling with such ideas is the aim of the next two Sections.

2 The Nature of the Deep Past

What is the nature of history and the deep past, and what difference might this make to our knowledge of it? As we've seen, Carol Cleland takes paradigm historical inferences to target token events, while experimental methods concern regularities, generalities and 'laws'. Her view echoes an older contrast between 'nomothetic' and 'idiographic' routes to knowledge. I'll start by asking whether historical science should be characterized in idiographic terms, as Cleland recommends. Answer? No: as we'll see, nomothetic and idiographic understanding are deeply intertwined – you don't get one without the other. However, this epistemic point leads to an ontological discussion: I'll highlight how some historical processes generate, surprising, specific and messy kinds; history often generates what I'll call *peculiarity*. This will set us up for Section 3, and our examination of historical explanation. To get there, I'll consider Collingwood's idea that history 'generates new forms', cash this out via a brief discussion of historical kinds, and then introduce peculiarity.

2.1 Idiographic and Nomothetic

The nineteenth-century philosopher Wilhelm Windelband introduced the terms *idiographic* and *nomothetic* to distinguish two routes to knowledge. An idiographic investigation takes the target on its own terms – it is treated as unique and its individual history is traced and narrativized. Think autobiography. A nomothetic investigation treats its target as a token of a type or an instantiation of a kind. Think particles acting in accordance with laws. In 1980 Stephen Jay Gould implied the distinction while discussing what would eventually be known as 'the paleobiological revolution'.

> Paleontology is, in large part, a historical discipline charged with documenting the irreversible and unrepeatable events of life's history ... We care very much that *Neotrigonia* lacks (except in its larval shell) the discrepant ornament that characterizes most mesozoic trigonians. The splinter that retarded the ball rolling down this particular inclined plane is merely a nuisance (Gould, 1980, p. 98).

Gould compares palaeontologists (who care about the particular morphology of extinct clams like *Neotrigonia*) and physicists (who care about how balls might roll down inclined planes). The former are interested in unique, often contingent events. As such, they emphasize features that make their target different. The latter are not interested in how a particular ball might find its way down a plane, but in the rules by which all ball-like objects behave. That is, *laws*. Another way of dividing idiographic from nomothetic, then, is simply to say that the nomothetic investigations care about and use laws of nature, whereas idiographic investigations don't.

So, do laws make the difference between historical and non-historical knowledge? Laws, I take it, are regularities, but not all regularities are laws. Laws are *non-accidental* regularities. By 'regularity', I just mean an event's recurrence. An *accidental* regularity is one that holds, but in a fragile way: it just so happens to hold, as a result of a fluke. *Non-accidental* regularities, by contrast, occur for a reason. The rules of taphonomy (the science of fossilization) are not accidental. That, for instance, hard-bodied animals are more likely to fossilize is a result of the nature of fossilization. Because fossilization requires that organisms maintain morphological integrity under pressure, squishier organisms are less likely to leave a mark on the fossil record. The bias towards hard-bodied critters isn't some fluke.

Why are non-accidental regularities important? Explanations often concern themselves with how things would have been different. Had the continents of the mid-Cretaceous not broken up, then perhaps the mid-Cretaceous revolution would not have occurred, say. As I'll suggest below, explanations explain by situating their targets in various ways – and often they are situated within spaces of possibility. Explanations show that had different conditions obtained, then a different result would have occurred. Non-accidental regularities are stable enough to support such counterfactuals: thus, they tell us what would have happened otherwise. On this approach, laws make scientific explanation possible. Further, much scientific knowledge is based on induction: a set of cases is examined, and we extrapolate to the general class. Schott et al. (2011). suggest that their study of two *Pachycephalosaur* species (or just one, if their ontogenetic hypothesis is right) gives reason to think that *Pachycephalosaur*s (or even dinosaurs) generally underwent surprising ontogenetic changes. Inductions take us from individuals to patterns. However, if the patterns are merely accidental – just happen to hold – then establishing the pattern in one set of cases is less likely to secure its occurrence in further cases. If the regularity isn't accidental, then inductive inferences are supported. Presumably, regularities support counterfactuals and inductions when there is some underlying reason for the regularity to hold. Dinosaurs shared developmental mechanisms, and it is these mechanisms that enable more-or-less extreme changes during ontogeny. This thought partly warrants Schott et al.'s inductive inference.

Understood as non-accidental regularities, laws play a role in any science. Laws, and theorizing about laws, set our expectations about what will occur, thus defining what is surprising, and making some forms of explanation possible. Some accounts of laws go further, claiming that laws tell us how things must be: describing necessities (Armstrong, 1983). But the cases driving our reflections thus far include examples of non-accidental regularities that hold under restricted conditions (sometimes called '*ceteris paribus*') and there are often exceptions within those conditions (Mitchell, 1997). Normally, hard-bodied critters are more likely to be fossilized than squishy critters. However, this trend may be bucked. The famous Burgess Shale biota date to the Cambrian (around 509 million years ago) and contain bountiful, diverse soft-bodied organisms (my favourite being the five-eyed, claw-at-the-end-of-a-trunk *Opabinia*). Such assemblages are more common than we might expect, roughly 40 locations being known (Gaines et al., 2008). These formations occur via a different-from-usual process of fossilization (although which process is still debated). Regardless, the 'hard-bodied-bias' in the fossil record is a non-accidental regularity (there is a good reason for it) despite having both systematic exceptions (when the underlying conditions are different) and particular, fluky exceptions where soft-bodied organisms get lucky in the fossilization lottery. This doesn't stop regularities from supporting counterfactuals or inductive generalizations, so I'm happy to consider them as being of a class with full-blown, necessary laws. I don't think there is a clear-cut answer as to how reliable a regularity must be to count as a law: I suspect this will depend on context.

With this account of 'regularity' on the table, we can reframe the nomothetic versus idiographic dichotomy. A nomothetic science is primarily interested in understanding regularities, whereas an idiographic science is interested in explaining particular events. Some philosophers have suggested historical science is idiographic in this sense. Take Carol Cleland: 'the hypotheses of prototypical historical science differs from those of classical experimental science insofar as they are concerned with event-tokens instead of regularities among event-types' (Cleland, 2002, p. 480).

On this view, the main target of historical science is singular events in our world's particular history. Ben Jeffares (2008) responds: 'the historical sciences are as interested in understanding the general causal structure of the world as much as any other branch of science' (Jeffares, 2008, p. 475). Jeffares' argument focuses on midrange theories. As he points out, such theories just are regularities – they are nomothetic. Fossilization is not a particular event, but an event-type (in fact a set of event-types). As such, paleontologists studying taphonomy are concerned with 'regularities among event-types'.

Cleland is dismissive: 'generalizations of this sort play a secondary role in historical research. They are not the targets of historical research but rather useful tools borrowed from other disciplines for special purposes' (Cleland, 2011, p. 566). This looks like an impasse: Jeffares (also Turner, 2013; Currie, 2018a, chapter 7) points out that regularities play a role in historical science, but Cleland relegates them to supporting roles, mere handmaidens to the real business of uncovering the past. How can we adjudicate? I think historical scientists are interested in both the idiographic and the nomothetic. This is because the general and the particular are not separate in practice. To understand the particular, we need to think of them as (at least partially) partaking in, and being shaped by, regularities; in understanding particulars we learn about regularities.

Recall Schott et al.'s analysis of the *Pachycephalosaur Stegocera validum* (2011). Although their work explicitly targets patterns of ontogeny in *Pachycephalosaur*, they focus on a careful analysis of two species. Indeed, their work was motivated by Goodwin et al.'s analysis of a single specimen. Goodwin et al. (1998), examining a single skull, noticed it was marked by highly vascularized and 'fast' growth patterns – suggesting that the dome grew later in ontogeny. As we've seen, Schott et al. examine a larger data set from multiple empirical perspectives while connecting their results to more general pictures of dinosaur diversity and ontogeny. They consider the idiographic and nomothetic together. In Schott et al.'s concluding discussion, they start with a lesson about the ontogeny of the *Pachycepholosaur* species they examine, and its relationship with one other putative species: '[we have shown] that as *Stegoceras* matured, the skull changed shape dramatically, and demonstrates conclusively that *Ornatotholus browni* represents a transient ontogenetic stage of *S. validum*'(p. 19).

They then link this discussion with the relationship between domed and flat-headed *Pacyhcephalosaur* generally: 'The extensive nature of these changes is such that juveniles and adults differ radically in their general appearance, and we hypothesize that this model of dome growth is a common developmental trajectory of domed *Pachycephalosaurs*' (p. 20).

Which is itself linked to yet broader discussions about ontogeny in some late-Cretaceous dinosaurs: 'The phenomenon of extreme morphological differences between juveniles and adults is becoming increasingly well-documented in ornithischian dinosaurs, including Hadrosaurs and ceratopsids' (p. 20).

Which matters – as we've seen – for explaining the possibly inflated biodiversity of the Cretaceous: 'Historically, these transitional juvenile morphologies have been erected as distinct taxa, resulting in artificially inflated estimates of biodiversity in these groups' (p. 20).

Estimates of biodiversity in late-Cretaceous dinosaurs are surprising in virtue of our general expectations about biodiversity in large-body critters, themselves grounded in examinations of biodiversity across a range of taxa. The nomothetic (vertebrate herbivores fall within particular ranges of biodiversity) and the idiographic (*Ornatotholus browni* are infant *Stegoceras validum*) are intimately related: you don't get one without the other. Notice the iterativity of the research. Theoretically motivated expectations about biodiversity motivate new field research, the results of which underwrite new reconstructions of the particular lineages of the late-Cretaceous, which are themselves informed by, and fuel for, new sets of expectations about biodiversity. Models of regularities are clarified and honed by interactions with particular cases.

The nomothetic sets research direction (general concerns about dinosaur ontogeny motivated Horner and Goodwin's fieldwork); by generating puzzles, the nomothetic sets what is significant about cases (dinosaur biodiversity in the late-Cretaceous is odd given general patterns in biodiversity); the nomothetic influences how traces are interpreted (theories of dinosaur ontogeny help determine when to classify specimens as different species or different growth-stages); the nomothetic provides grounds for linking or associating cases (theories of dinosaur phylogeny highlight the continuities between *Pachycephalosaur* and *Triceratops*). The idiographic provides empirical support and tests for theories (Horner and Goodwin's *Triceratops* analysis suggests that dinosaurs had unusual ontogenies); the idiographic inspires hypotheses (Goodwin et al.'s analysis of a single specimen suggested that flat-headed *Pachycephalosaur* represented an early developmental stage); the idiographic provides concrete examples of what would otherwise be merely abstract and theoretical (Horner and Goodwin's ontogenetic sequence is a striking example of an unusual ontogeny); the idiographic generates surprises (why would the emergence of flowering plants have such a profound effect?).

The historical sciences are both nomothetic and idiographic. And surely so are most sciences. But this is not the end of the story. Although historical sciences care about regularities, perhaps they are regularities of a different sort to those other scientists care about. That is, perhaps the nature of history is such that regularities take a particular form. Before considering this possibility, a brief digression is required.

2.2 A Historiographical Tangent

A view about the relationship between history and science clamours for attention. Adrian! (the thought goes) you've made a fundamental error by lumping history together with natural sciences. Human history is special because it has humans in it: intentional beings. History really matters when it is human history.

The clearest exponent of such a perspective is Collingwood (1976/1936, see also White, 1966): 'so far as our scientific and historical knowledge goes, the processes of events which constitute the world of nature are altogether different in kind from the processes of thought which constitute the world of history' (p. 170). Why draw so sharp a boundary between processes of nature and processes of history?

Collingwood considers two ways of undermining analogies between natural science and human history. First, he points out that historical kinds are not eternal, but emerge over history:

> Change and history are not at all the same. According to this old-established conception, the specific forms of natural things constitute a changeless repertory of fixed types, and the process of nature is a process by which instances of these forms . . . come into existence and pass out of it again . . . Now in human affairs . . . there is no such fixed repertory of specific forms. Here the process of becoming was already by that time recognised as involving not only the instances or quasi-instances of the forms, but the forms themselves (p. 166).

The idea is this: in the natural world, fixed forms are sometimes instantiated, sometimes not; in the world of history, the forms themselves arise. Sometimes a single hydrogen and two oxygen molecules form a chemical bond and become water, and then sometimes that bond is broken. But this is merely change, not history. Historical change involves new forms arising. Human institutions do not exist time immemorial merely to be instantiated, but are created in particular times and places. The European Union was not merely instantiated, but was founded with the signing of the Maastricht Treaty in 1992. And the particular shape, properties – and fate – of that institution are intimately linked to those historical conditions.

However, Collingwood points out that this cannot divide nature from history, because sciences like biology and geology have new forms aplenty. '*Pachycephalosaur*' is no occasionally instantiated eternal kind. It was a new form, arising in the revolutions of the mid-Cretaceous. In fact the whole natural world looks historical in this sense: 'Today even the stars are divided into kinds which can be described as older and younger ... the chemical composition of our present world is only a phase in a process leading from a very different past to a very different future' (p. 167).

So, what then is the difference between history and nature? Collingwood's second pass places the difference in intentionality:

> The archaeologist's use of his stratified relics depends on his conceiving them as artefacts serving human purposes and thus expressing a particular way in which men have thought about their own life; and from this point of view the palaeontologist, arranging his fossils in a time-series, is not working as an historian, but only as a scientist thinking in a way which can at most be described as quasi-historical (p. 168).

Unlike the scientist, the historian and archaeologist have to get into the minds of their subjects. Collingwood distinguishes between the 'outside': the world of events and causes, and the 'inside': the world of intentions, beliefs and desires. Approaching some event, the scientist links it to other events and laws: *situating* it (as I'll say later) in the world's causal flow. Approaching some event, the historian moves from the outside to the inside: to the intentions of the historical actors. How do historians do this? By applying their own agency to it, by 'rethinking them in his own mind' (p. 169). And this action of rethinking is central to the reflexive nature of historical practice: 'The historian not only re-enacts past thought, he re-enacts it in the context of his own knowledge and therefore, in re-enacting it, criticizes it, forms his own judgment of its value, corrects whatever errors he can discern in it' (p. 169). In this sense, the process of history is a kind of dialogue between the historian and the past actors.

I think Collingwood has insights concerning the value of history, which I'll return to in my conclusion. But, has Collingwood provided a convincing way of drawing the historian and scientist apart? In one sense, perhaps, but in another no. We should disambiguate two related claims in Collingwood: (1) history requires psychological explanation, (2) psychological explanations necessitate 're-enacting'. The first claim might well be true, but unless there is something special about psychological explanation, this doesn't tell us anything about the relationship between history and knowledge. The second suggests what might be special about psychological explanations: they require the historian 'rethinking' the thoughts of past actors. But I think this underemphasizes the role 'rethinking' plays in natural science.

Science is not just about the external. To say it is suggests that the totality of scientific knowledge is *explicit* and *propositional* – that scientific knowledge doesn't involve 're-enacting'. I think this misconstrues some (if not all) scientific knowledge: there is good reason to see science as a tacit, skilled, communal, human endeavour. As this thought has a fair pedigree in philosophy of science, let's take an idiosyncratic tour.

We'll start with Thomas Kuhn (1970) who I think gets it right when he emphasizes the importance of institutional pedagogy in shaping (and making possible) scientific knowledge (see also Polanyi, 1958):

> [scientific knowledge] is not acquired by exclusively verbal means. Rather it comes as one is given words together with concrete examples of how they function in use; nature and words are learned together . . . what results from this process is 'tacit knowledge' which is learned by doing science rather than by acquiring rules for doing it (Kuhn, 1970, p. 191).

The process of becoming a scientist involves acquiring a diverse range of knowledge, skills, behaviours and values. Vertebrate paleontologists know an enormous amount about animal anatomy; they know how to do productive fieldwork, they know how to use various experimental and statistical tools. In my experience paleontologists are excellent at detecting and processing visual information, presumably because they spend so long on fieldwork and identifying finer points of their chosen taxa's anatomy. Vertebrate paleontologists have a sense of what are good questions and what are good answers in the context of their science. They have what Kirsten Walsh and I have called 'explanatory expectations' (Currie & Walsh, forthcoming). That is, there are particular explanatory forms that they have learned to deem satisfying. Kuhn's point is that this knowledge is not simply told, but is imparted and demonstrated in often tacit ways in lecture theatres, laboratories and in the field. Much of what it is to be a scientist is (1) tacit and (2) imparted through a range of formal and informal pedagogical processes.

One way of capturing this aspect of Kuhn's work is via Nancy Nersessian's 'mental models'. Nersessian aims to explain mechanisms of conceptual change

(1999, 2007) by positing representations in scientists' minds that act as a kind of frame through which their scientific work is done. 'Concepts provide a means through which humans make sense of the world. In categorizing experiences we sort phenomena, noting relationships, differences, and interconnections among them. A conceptual structure is a way of systematizing, of putting concepts in relation to one another in at least a semi – or locally – coherent manner' (Nersessian, 2007, p. 126).

Because conceptual structures are large, complex – logical – beasts, scientists require mental models to understand them. These models are partial, good-enough-for-jazz representations of conceptual structures; they are incomplete and often non-explicit. Nersessian employs the notion of mental models to explain conceptual shifts in science. For our purposes, the point is that at least some scientific knowledge is constituted by partial, incomplete, often idiosyncratic representations, which scientists use to make sense of theories and engage in collaborative work. Although the target is not typically imagined as literally intentional, I think the internal know-how of Nerssesian's mental models bears a striking resemblance to Collingwood's re-enacting. Both involve employing internal, often tacit, cognitive tools to get a grip on some phenomena of interest.

This tradition of emphasizing the tacit in science has recently expanded to an emphasis on *understanding* as an epistemic good, differing from explanation or truth. These approaches undermine Collingwood's reliance on scientific knowledge being 'external' in the sense of being explicit. In one recent example, Angela Potochnik has argued that the ultimate aim of science is not explicitly truth, but rather 'understanding' (Potochnik, 2017. For her, 'understanding' has two components: first, a 'sufficiently true' component: if I understand some aspect of the world I need not know it all that precisely or carefully, but I'd better get some things right; second, a psychological 'grasping' condition: the understander has to 'get it' in some way. She explains many aspects of scientific work, the ubiquity of idealization especially, by claiming that science isn't after truth but is 'responsible to and reflective of human particularities' (p. 199). That is, our limited and idiosyncratic needs and interests as human knowers. Other philosophers have emphasized the roles various kinds of know-how and abilities play in science (Le Bihan, 2016; Leonelli, 2016). The more we humanize and socialize scientific knowledge – think of it as a product of the epistemic activities of social beings like us – the more we emphasize the non-explicit aspects of it and, I think, the less convincing we should find Collingwood's attempt to split science and history in terms of method.

Let's return to Collingwood's comparison of the archaeologist and the paleontologist. Although paleontological practice doesn't rely on conceiving the materials as human artefacts, if the accounts of scientific knowledge we've considered are right – and I think they are – then scientific knowledge is highly internal, often involving 're-enactment' (through something like Nersessian's mental models) in the context of scientists' own knowledge. There is plenty of the internal in science, then. You might insist that employing tacit resources to re-enact a psychological explanation is fundamentally different from using them to understand the 'external' events of scientific explanation, but this insistence needs backing up: in virtue of what, if not requiring re-enactment, might intentionality be a fundamentally different epistemic target?

2.3 Kinds

I think Collingwood got it wrong when he distinguished history (and archaeology) from natural science, but I think he had his finger on the button when he linked the emergence of 'new forms' to history. In the next two Sections I aim to make sense of this idea. Most pressingly, what are we to make of Collingwood's rather metaphysically weighty, Platonic sounding appeal to 'forms'? I think reflection on historical kindhood can provide a less off-putting (for those who are off-put!) account of what is being got at.

There are many ways of carving up the world. We could do it via, say, height, or colour, or age, or mass, or alphabetically, and so on (ad infinitum). The notion of 'kinds' suggests that some ways of carving are special: they reflect some natural order, perhaps. At a first pass, the kinds that interest scientists – say, species or genes in biology, layers of strata in geology – seem less arbitrary than categories like 'blue things' or 'things I like'. Analyses of 'natural kinds' often track this intuition (Hawley & Bird, 2011). On some accounts, natural kinds pick out a privileged, hierarchical order: they 'carve nature at the joints'. On another extreme, there is no real difference between kinds like 'hydrogen' and kinds like 'things Adrian likes': all categorizations are subjective and equally legitimate. I fall somewhere in between. Humans draw many categories, there are many different 'kindings' in our practices, and these categories often do important work for us. Paleontologists divide fossil specimens into morpho-types based on diagnostic features. They then consider further categorizations which explain these morphotypes: sometimes into phylogenetic groupings such as species or genera, sometimes into intra-lineage categories like sexes or different life-stages. And these kinds do work for us. The ancestral divisions of phylogenies, for instance, matter for determining biodiversity, which as we've seen itself feeds into an array of further questions. Dinosaur diversity at the end-Cretaceous matters for understanding the K-Pg extinction event. Although many of these kinds are varyingly fuzzy – vagueness and ambiguity abound – I'm not inclined to draw a strict distinction between 'real' kinds and those which scientists merely project. The world is a complex place, and I'd be astonished if some small, prearranged set of categories uniquely pick out its structures. Indeed, as we'll see in the next Section, some historical processes should lead us to expect a wide diversity of kinds. But this doesn't mean that all kinds are equal: a kind's legitimacy is founded in the work it does for us, and we can't just force the world into whichever shape we want.

I think there is something to Collingwood's idea that in history 'new forms' – that is, new kinds – arise, whereas in other contexts there is a 'changeless repository of fixed types'. We needn't grant this difference too much metaphysical baggage, we can instead understand the difference in terms of different ways of kinding.

Laura Franklin-Hall (in prep) has distinguished between three types of kinds, each differing in how they lump and split tokens. First, *synchronic kinds*, which we might align with Collingwood's 'changeless fixed types'. These lump and split on the basis of the tokens sharing intrinsic – often essential – properties. Consider 'hydrogen'. An atom is hydrogen just in case it has a single proton in its nucleus. I needn't know anything about an atom's history to classify it as hydrogen. Franklin-Hall's next two kinds are both *diachronic,* where kind-membership depends in part on the token's history. Second, then, we have *type-historical kinds*. These categorize objects in terms of the processes that produce them. To be a member of the kind, a certain type of history is required. Although geologists might identify, say, igneous rock using morphology, what the kind picks out (what makes igneous rock igneous) is being formed via volcanic processes: being the remains of cooled lava. Third, we have (a bit ironically) *token-historical-kinds*. These, in a sense, are not 'kinds' at all, but rather pick out particular sequences or objects across time. You likely consider yourself to be a token-historical-kind: different time-slices of Adrian are 'Adrian' in virtue of being the same token-historical kind. A common example from the philosophy of biology is 'species': what makes you and I the same species is not our having the same set of essential properties (being 'featherless bipeds' perhaps), but that we are linked through ancestry (Hull, 1976). Species are not type-historical kinds but token-historical kinds. Species membership is decided on the basis of being part of the same token process. Being an igneous rock doesn't depend on forming from lava from the same volcano: hardened lava from Martian volcanoes would still be igneous. Not so for species membership: if the featherless bipeds are not ancestrally linked, they ain't the same species.

I think Franklin-Hall's distinction goes some way towards capturing Collingwood's insight. Where scientists traditionally did their 'kinding' with synchronic, essential kinds – where forms are in some sense merely instantiated – they now tend to kind historically. They do this because as history progresses, new kinds arise: volcanoes are required for igneous rock, for instance, and the kind 'igneous rock' has a history linked to the emergence of volcanoes.

Kindhood is nested and cross-cutting. Super-Paw is both a member of the token historical kind *Felis catus*, as he is of the *Carnivora* order and perhaps the synchronic kind *terrestrial, night-time predator* and the type-historical kind *living organism*. Which kinds are 'more legitimate' – where Super-Paw's kind-joints are – strikes me as a less important question than noting that such diverse ways of kinding are an integral part of many scientific practices. In the next Section, I'll suggest an explanation for this diversity, and for the power of diachronic kinds in historical science and elsewhere: what I call history's 'peculiarity'.

2.4 Peculiarity

I've been trying to make sense of Collingwood's idea that history creates new forms, which I'm understanding in terms of diachronic kinds. But why would attending to history lead to kinding in this way? Objects and processes have pasts: even if atoms may be categorized synchronically, individual atoms have pasts that might interest us. However, those pasts don't appear to make a difference to what makes them atoms. When, then, does history matter for categorization, when do diachronic kinds come into their own? The remainder of this Element is, in part, an argument that a feature of historical processes, their 'peculiarity', explains both the power of diachronic kinding and the prominence of narrative in historical disciplines. Why the term 'peculiarity'? This is in part a matter of product differentiation; it is simply a label. What I mean by 'peculiarity' is precisely the definition I provide and unpack below. Hopefully more casual understandings of 'peculiarity' aren't too offended by my term of art.

'Peculiarity' is related to philosophical discussions about contingency. Historical investigations are often concerned with events that, in some suitably strong sense, could have been otherwise. The mid-Cretaceous revolution wasn't the inevitable result of time's pre-determined unfolding; *Pachycephalosaur* headgear was the outcome of a specific, perhaps path-dependent route through a bewildering possibility space. Collingwood's idea that history generates new forms is related to historical contingency: in order for some things to happen, others must happen first (McConwell & Currie, 2017; Beatty, 2006; Ereshefsky, 2014; Desjardins, 2011 Powell & Mariscal, 2014).

To develop my account, I'll draw on recent discussions of contingency by Alison McConwell and Kim Sterelny. I'll start with McConwell (2019) who, building on John Beatty (1994), argues that contingency is generative. 'Contingency, by causing diversity of a certain sort, results in a pluralism. Specifically, contingency causes structural varieties, which distinguish a plurality of types' (McConwell, 2019).[3]

Let's unpack. We've seen many kinds of paleobiological diversity: in morphology, life-stages, phylogeny, biodiversity and so on. In biology, there are many diversity concepts and measures (Maclauren & Sterelny, 2008). McConwell adopts a pluralistic, interest-based conception of diversity. For her, determining 'whether there is diversity or not will depend upon the measurement strategies and tools'. She focuses on 'structural' diversity, which is contrasted with 'functional' diversity. One might group together, say, deer antlers and *Pachycephalosaur* skulls as being sexually selected for male-male competition. But the two differ structurally: the male-male competition is instantiated in different ways. Structural description ranges from being very coarse-grained, such as the four-limbed bilateral symmetry of vertebrates or the organization of ecosystems, to finer-grained cellular and biomechanical analyses.

McConwell's view takes evolutionary contingency to be the tendency of evolutionary processes to lead to highly particular outcomes. That is, the outcomes of evolution are highly sensitive to initial conditions or path-dependent. She argues that this feature leads to structural diversity. Evolutionary processes work on variation within populations: each individual's developmental suite is a little different from their conspecifics', and there is (sometimes bountiful) wiggle-room provisioned by plasticity and mutation. The range of actual and potential variation – which evolutionary paths are open and which are closed – are the product in part of these structural features, and are themselves the product of a long evolution. This, McConwell points out, leads to a more generative, path-dependent conception of evolution. '[O]ne can take this a step further: evolution is transformative, less optimal, and less constrained' (McConwell, 2019).

The breaking up of Gondwana likely set the stage for the modern world as new niches arose, and new critters evolved in response. By taking one path, some possibilities opened and others closed. Further, new trajectories were set: the insect-angiosperm alliance, and perhaps the eventual dinosaur extinction. By generating diversity, contingency necessitates pluralistic – and historical – kinding. Historical-token and -type kinds are appropriate because of these

---

[3] You might balk at McConwell's referring to contingency as a cause (she discusses this at length). For my purposes, it is not a worry whether we want to call an event's contingency, or the processes which underwrite the event's contingency, a cause.

historical processes. Evolution and these structural features themselves are, then, generative of 'new forms', we should expect a bounty of historical kinds.

Relatedly, Bill Wimsatt emphasizes how some processes (particularly evolutionary ones) often generate arbitrary or accidental features, which then become 'locked in' as essential parts of the system downstream. '[W]ith accumulating dependencies, seemingly arbitrary contingencies can become profoundly necessary, acting as generative structural elements for other contingencies added later' (Wimsatt, 2007, p. 135). The conversion of contingency into necessity is one way in which history makes her own rules: again, harkening to Collingwood's 'new forms'. Dinosaurs happened not to be well-suited for angiosperm consumption, but the flexibility of angiosperms in pollination – fruit and flowers – made them ideal for fast-evolving insects. This contingency became locked in as the insect-angiosperm alliance, which has been stable for a hundred million years. Similarly, Inkpen and Turner (2012) suggest considering contingency in terms of a topographical landscape that evolves over time. As time passes, what is likely or unlikely to occur can change. Wimsatt describes a process of an at-first contingent event becoming increasingly necessary over time but events also might become less likely. The overall lesson here is that various processes are more likely to produce diversity, contingency, necessity or homogeneity over time.

Kim Sterelny (2016) is interested in laws in human history. Instead of quibbling over whether such things exist, he asks why human history has both robust and fragile outcomes. Fragile outcomes are beloved by historians and story-tellers alike: if only it hadn't rained that day, if that particular cavalry charge had gone differently, had I turned left rather than right, then things would be different – and perhaps different in unfathomable ways. However, Sterelny points to robust patterns and outcomes in human history. In historical science we also see both fragility and robustness.

As I've mentioned, one reason why we want an accurate measure of late-Cretaceous dinosaur diversity is to develop explanations of the K-Pg extinction. Although models of the asteroid impact explain the massive biodiversity plummet in the oceans (mostly through increased oceanic acidity), it is unclear why the same occurred on land. One hypothesis blames a decrease in ecological resilience among larger dinosaurs (Mitchell et al., 2012),in particular, a decrease in the diversity of large herbivores towards the Cretaceous' close. Ecological modelling suggests that large herbivores like *Triceratops* play important roles in ecosystems. Their young and eggs provide food for smaller carnivores, adults are prey for larger predators, and their remains and dung fertilize soil. Because they are hooked into ecosystems at multiple levels, disrupting these herbivores can have disastrous consequences. On this scenario, towards the end of the Cretaceous, although total populations didn't decrease, large herbivore biodiversity did (they became more homogenous), leaving the ecosystem dependent on the fate of a few species. Under these conditions – perhaps – the dinosaur's fall was foreordained: critically weakened ecosystems were simply waiting for the right trigger. When exactly the fall would happen, or what the trigger would be, might be unknowable before the event, but the collapse being triggered was extremely likely.

So, history contains both fragile – contingent – and robust events. Why?

Sterelny's answer is that some historical outcomes and trajectories depend 'only on aggregate effects of interactions in populations' (p. 532). In the scenario above, the actions or fate of individual *Triceratops* don't make a difference: the ecosystem's collective instability is to blame for the collapse. This is what underwrote the extinction's robustness. Further, Sterelny notes that some systems have stabilizing mechanisms. Developmental and inheritance systems in biological lineages, for instance, ensure that information is carried pretty robustly across generations. In human systems, however, structures sometimes emerge that actively work against aggregate effects quashing a particular individual's contributions. Sterelny calls these 'command structures': they arise when communities shift from egalitarian modes to hierarchies with some individuals in charge: chiefs, generals, monarchs and bosses: '[command structures] are a source of contingency for two reasons. They make population-level trajectories sensitive to the actions of the few, and they make that few more likely to behave in erratic, unpredictable, hair-trigger ways' (p. 534).

Ironically, the rather robust result that human societies develop hierarchical control as populations increase sometimes also produces fragility: the whims of Dear Leader.

McConwell and Sterelny's insights can, I think, be synthesized into a story about just when and why history matters to knowledge. McConwell emphasizes how contingent processes can be generative, acting as a scaffold for new structures. Sterelny points out that the robustness or fragility of trajectories depends in part on the structures of those trajectories. McConwell: history generates a diversity of structures; Sterelny: robustness and fragility depends upon structure. Bringing these together generates what I'll call *peculiarity*. Before introducing the account abstractly, let's consider a toy case.

Imagine a factory that produces widgets. During production, the factory has a set of manipulable processes, which determine the widget's properties. For instance, we might include quality-control staff to ensure the widget's homogeneity, or we might include more colours in the production process, leading to differently coloured widgets. This latter could be randomized, for instance, leading to widgets of a wide variety of colours. On leaving the factory, the widgets will be differently attractive to consumers depending on the widgets' properties and consumer preferences. It may be that consumers desire reliability, at which point homogeneity-producing processes will be valued and reinforced. Or they may desire diversity, thus introducing more colours would boost sales. In this scenario we have (1) a set of processes (how the factory produces the widgets), which effect (2) the properties of the outputs (the colour of the widgets, say), which (3) make a difference to the desirability of the widgets (if customers prefer diversity, say), and thus the factory's success. Under those conditions, there is a dependency between the processes that produce widgets and the widget's subsequent success. This needn't be the case: consumers might be indifferent to diversity and only care about reliability, at which point the processes producing a variety of colour make no difference to widget desirability.

Note the possibility of feedback between the factory's processes, the properties of the widgets and public desirability. It may be that some consumers are 'stamp-collectors', that is, they value off-kilter, rare, variations in an otherwise homogenous set. If we introduce sources of error or otherwise variation-introducing features into the factory's processes, this might lead to stamp-collectors becoming interested in widgets. This would encourage the factory's processes to generate a homogenous – but not too homogenous – output. Here, a particular dynamic in the factory's processes has emerged in light of the interaction between widget-properties and the stamp-collectors.

This toy case can be used to illustrate the relationship between the insights of Sterelny and McConwell. McConwell describes how different processes are generative: a factory with some error in production potentially makes room for new dynamics to occur. Sterelny describes how the robustness or fragility of a system depends, in part, on its properties. Depending on the consumers, a factory with reliable widgets will have a fate different from one with less reliable widgets. Putting these together, we have a dependency between a set of generating and maintaining processes (the factory making widgets) and a system's properties (the widget's properties and their subsequent desirability). Such dependencies are the basis of *peculiarity*, a fairly ubiquitous relational property, which, nonetheless, can be used to capture how history matters to knowledge. Here is the notion abstractly:

> Some target is *peculiar* to the extent that its modal profile is sensitive to the properties of the processes that generate or maintain it.

Let's break this down. First, we have some *target*. I'm going to be very inclusive about what counts as a target: the entire Cretaceous period, the

terrestrial revolution, the fate of a particular *Triceratops* . . ., anything we might
be interested in explaining can be a target. Second, targets have a *modal profile*.
To a first approximation, we might equate modal profiles with a kind of
robustness: the likelihood of the target's properties staying the same or changing. Likelihoods are conditional probabilities. Given certain conditions,
changes and perturbations in background conditions, how likely is the target
to stay the same or change? But a target's modal profile isn't simply the
probability of it changing: modal profiles include the ways a target may trans-
form under various conditions. There might be conditions under which the
angiosperm-insect alliance would become less resilient (perhaps if there were
a precipitous drop in insect populations . . .), or chances that its dynamics might
shift (becoming more dependent upon artificial fertilization, say). A modal
profile, then, isn't simply robustness, but a map of the various conditions
under which a target might change.

As we've seen, Sterelny emphasizes how different structures are more-or-
less robust, and are thus an important part of a target's modal profile. Structures
whose behaviours are aggregate effects tend to have robust outcomes (although
may not be robust themselves: the dinosaur extinction may be a robust outcome
even if – actually because – dinosaur ecosystems were not themselves resilient).
Other structures increase contingency, such as some hierarchical control structures. But to generalize from Sterelny, function can affect modal profile as well.
The modal profile of *Pachycephalosaurus* depends, in part, on the function of its
thickened skull. If the function is mate-mate competition, or if it is display, the
subsequent evolutionary profiles are different. For instance, if the former holds,
the direction of evolution depends on other males, whereas if the latter holds, it
depends on female mating preference. By 'modal profile', then, I mean the
kinds of changes and transformations (and the accompanying likelihoods)
a system may undergo given the system's properties and relevant background.

Third, targets arise from, and are maintained by, *processes*. As with targets,
we should be open-minded about what will count as a process. The splitting of
Gondwana was a process that potentially led to the emergence of the insect-
angiosperm alliance (and the undermining of dinosaur dominance); but
Gondwana's unity prior to the terrestrial revolution was a process that maintained the dinosaur-gymnosperm alliance of the Jurassic and early-Cretaceous.
When peculiarity holds, the target's modal profile is sensitive to some features
of these processes. What kinds of features? One way of capturing the relevant
features is to ask whether the process is *diversity-boosting* or *dampening*. In our
factory, we might change the properties of the widgets, or the consumer's
preferences regarding the widgets, either way these change the likelihood of
the target changing or remaining the same over time. So, a process is diversity-

boosting when, versus some competitors, it is more likely to generate a diverse set of outcomes. Because they encourage population isolation and environmental variability, dispersed continents are more diversity-boosting than a mega-continent. Often, extremely stable systems are such, in virtue of dampening diversity. A too-diversity-boosting developmental system (one with a high mutation rate for instance) is unlikely to generate a viable organism. So by dampening diversity, stability is achieved. Diversity-dampening or boosting aren't the only features of processes that might matter, but they're prominent in the cases we're considering.

Peculiarity does not turn on the robustness or fragility of the target, but on the extent to which the target's modal features result from structures which are themselves the outcomes of generating or maintaining processes (if you want, they are second-order fragile, even if they are first-order robust). Returning to our factory, diverse or homogenous widgets might be produced, but that does not make these peculiar. A process aiming for heterogenous (or homogenous) products – if it succeeds – might produce peculiarity if that is desired or disliked by consumers in a fashion that makes a difference to whether the factory continues producing widgets of that type or not. Peculiarity requires that the generating or maintaining process' features are responsible for the modal profile of the target. That is, it captures a sensitivity between, on the one hand, a generating or maintaining process and, on the other hand, the modal profile of the target. If I change the properties of the process (making it more diversity-boosting, say), does this change the target's modal properties? If so, then the target is peculiar regarding that process. The extent of peculiarity depends on how much the modal profile of the target changes given that counterfactual.

Peculiarity is sensitive to description; how something is peculiar depends, in part, on our focus. If all dinosaurs share the same developmental systems, then, in part, the surprising *Pachycephalosaur* ontogeny is maintained by diversity-dampening inheritance processes that ensure continuity across dinosaurs. Qua dinosaurs, the distinctive ontogenetic patterns they exhibit could be a result of processes occurring after dinosaurs diverged from crocodilians. In light of this sensitivity, it might make more sense to ask whether a target is peculiar in relation to some set of processes than to ask whether a target is peculiar per se. Presumably, for almost every target there will be some process to which they are relevantly sensitive, but still some modal profiles exhibit less constraint from those processes than others and, as I'll argue, it is for those more constrained, more sensitive targets that history matters.

Let's consider the peculiarity of dinosaur morphological development under the assumption that Horner and company's hypothesis, that they exhibit wide variation across ontogeny, is true. First, living relatives can help determine to what extent this feature of their development depended on past processes. Birds and crocodiles exhibit variation over their lifetimes, crocodilians especially often adopting different behaviours and ecological roles as they age, but we see nothing like the morphological variation expressed in *Triceratops* or *Pachycephalosaurus*. No anatomist worth their crust would mistake a baby croc for a new kind of croc. This suggests that dinosaur development was caused by changes occurring post their divergence from birds and crocodiles, for if it weren't, then we'd expect patterns of development to be stable, conserved, across these taxa. A non-generative, non-contingent process wouldn't produce diversity. Does the modal profile of dinosaur ontogeny depend on structures which are the product of those processes? Yes: the developmental systems underwriting the radical ontogenetic diversity of dinosaurs is bequeathed in large part from their evolutionary history. Dinosaur ontogeny, then, is peculiar: it is well-captured by diachronic kinding.

Recall the aim of this part: to make some headway on the nature of history. We started with Collingwood's intriguing suggestion that history generates new forms. I then suggested that we can understand these new forms in terms of diachronic kinds: categories where membership turns on having a particular history. Peculiarity, I think, leads to diachronic kinds. Peculiarity is an inherently historical notion, because it involves earlier processes providing the conditions that give rise to later states of affairs. But it doesn't capture just *any* temporally extended process: it is restricted to those targets who have developed, or emerged as a result of, particular dependencies with generating or maintaining processes. Such processes often provide a basis for Collingwood's 'new forms'. The angiosperm-insect alliance that dominated the Cenozoic has been highly stable. And it is plausibly highly peculiar, if indeed it depended on the continental break-up and on the emergence of the highly diverse, plastic developmental systems of angiosperms. Peculiarity, then, is not merely temporal, but could reasonably be called historical.

Merely being past is not enough to be historical, and I suggest that peculiarity (or relative peculiarity) can tell the difference. I suspect peculiarity is a near-ubiquitous feature of our world: look hard enough at anything and you'll find some peculiarity. History's reach is long and the popularity of diachronic kinds is a testament to it. To what extent peculiarity occurs is an open empirical question. However, my argument doesn't quite turn on peculiarity's commonality, I'm making a conceptual claim: history matters under conditions of peculiarity. As some targets are more peculiar than others – their modal profiles more dependent on particular processes – we should expect these highly peculiar targets to, as it were, be more demanding of an historical treatment (we'll get to those historical treatments in the next Section).

Accounts of historical or evolutionary contingency often focus on historical outcomes' sensitivity to initial conditions and path-dependence (Beatty, 2006; Desjardins, 2011). On such views the fragility and specificity of history's path are emphasized. I think these are inadequate for understanding how history matters for two reasons. First, they have difficulty capturing the remarkable stability of many of history's players. The dinosaurs dominated Earth's ecosystems for over a hundred-million years – they were very robust – but this doesn't make them any less historical. Second, by focusing on outcomes, they de-emphasize the importance of the processes that enable and shape history and their capacity to generate new kinds and dynamics (McConwell & Currie, 2017). Peculiarity is richer than these accounts and, as we'll see below, is better placed for understanding how history matters for knowledge.

History's peculiarity goes some way, I think, to explaining a major theme of Section 1: the localness of justification and knowledge in historical science. I argued that abstract, coarse-grained ways of thinking about evidence don't gain purchase because the licence of historical inferences typically depends on particular and idiosyncratic factors. And this is just what we should expect from peculiarity. As peculiarity sometimes involves new forms, new dynamics and new diachronic kinds arising, so also will knowledge of such systems be increasingly restricted to those local conditions. The idiosyncratic, opportunistic methodology of historical scientists (and, I imagine, many scientists overall) is an adaptation to the peculiarity of their subject matter.

Why does peculiarity matter for questions relating to history and knowledge? I'm going to argue that history matters for knowledge when history is peculiar. To demonstrate this, I'll turn to historical explanation.

3 Historical Explanation

History raises bountiful *whats* and *whens*: the mid-Cretaceous terrestrial revolution saw increased biodiversity in many land-based lineages; it occurred roughly 125 million years ago. But history also provides *hows* and *whys*: the emergence of flowering plants opened new niches for insects to occupy, thus perhaps partly driving rapid diversification. Philosophers have long worried about the relationship between *descriptions* – whats and whens – and *explanations* – hows and whys. What's the difference between describing and explaining? In virtue of what does explanation bring more to the table than mere description? In this part, I'm interested in whether history makes a difference to explanation. I think it does: specifically, history's peculiarity makes a difference to the appropriateness of various explanatory forms.

I'll start by discussing narrative explanation in Section 3.1. In Section 3.2 I'll provide a thin, ecumenical account of the difference between descriptions and explanations: I'll argue that explanations are descriptions which are situated in a way that generates understanding (or other explanatory goods). In Section 3.3, I'll argue that peculiarity is well suited to particular kinds of situating: narrative ones. As such, we can see why narrative forms of explanation are so suitable to history – because they capture peculiarity.

How general is my account? I'd love for it to turn out that I've captured every instance where history matters for knowledge. However, my discussion is no doubt driven by the examples we've been considering: macroevolutionary change and paleobiological studies of ontogeny. It may be that other contexts exhibit different kinds of historicity, but this would need to be shown. So, I'm officially agnostic about my view's scope: perhaps take it to be a speculative hypothesis aimed at opening discussion. I'm perfectly happy if my view ends up being an important part – if not the whole story – of the relationship between history and explanation.

3.1 Narrative Explanation

Historical explanations seem to take a narrative form. Consider: The modern world was formed in the mid-Cretaceous. Crucial aspects of modern biota emerged then, as did familiar geography. The mid-Cretaceous terrestrial revolution occurred during the super-continent Gondwana's final break-up, and these are plausibly linked. Super-continents are diversity-dampening. A large single landmass leads to comparatively homogenous environments, and isolated populations are rare. The emergence of modern continents encouraged heterogeneous biomes, founder populations and so forth. Another of the revolution's important ingredients was the radiation of angiosperms. This led to a whole new host of mutualistic alliances between flowering plants and their pollinators, which in turn scaffolded new niches: consuming angiosperms or predating on their pollinators.

The success of angiosperms – their multiple radiations, their very high biodiversity (a quarter of a million species!), and so forth – has received multiple explanations. As Benton (2010) puts it, their success was, in part, a result of key innovations coupled with high morphological plasticity. Angiosperms developed edible fruit, rapid growth and efficient hydraulic systems. Benton argues that angiosperms drove insect diversification, not vice versa.

> Studies so far suggest that the correlates of angiosperm diversification are manifold, and that insect pollination, for example, is not an adequate driver on its own. Combined fossil and molecular phylogenetic data show repeated

bursts of diversification, associated with the evolutionary introduction of novel functional traits throughout the history of the flowering plants (Benton, 2010, p. 3672, references removed).

It has been suggested that dinosaurs benefitted only minimally from these new angiosperm niches: evidence that they ate angiosperms is equivocal (Ghosh et al., 2003). As the team who introduced the notion of the mid-Cretaceous terrestrial revolution put it:

> Plant-eating insects and mammals very likely benefited more from the new sources of plant food. Detailed studies of dinosaurian herbivory and plant evolution has already suggested there was limited evidence that angiosperm diversification drove the Cretaceous diversification of dinosaurs. Our new evidence confirms that the [mid-Cretaceous terrestrial revolution] was key in the origination of modern continental ecosystems, but that the dinosaurs were not a part of it. Hadrosaurs and ceratopsians showed late diversifications, but not enough to save the dinosaur dynasty from its fate (Paleobiology research group, accessed 23/11/2018, references removed).

A picture emerges: the break-up of Gondwana and the evolution of angiosperms created a wide array of new niches that drove diversification in a range of terrestrial vertebrates. Dinosaurs, perhaps as a result of over-specialization on gymnosperms and ferns, were slow or limited in their ability to capitalize on these new niches. This shifted the 'world order', which had held from the late Triassic to the early-Cretaceous: an alliance between gymnosperms and dinosaurs on relatively homogenous super-continents; thus fatally undermining the basis of dinosaur dominance.

I'm less interested here in the evidential support this explanation may or may not have than I am in the explanation's form: what we have here looks like a narrative. In this narrative, we come to understand the emergence of the modern world by telling a story about how those features emerged. At base, a narrative explains some outcome by identifying a sequence of events – a trajectory – involving that outcome, often culminating in the explanatory target (Currie, 2014; Currie & Sterelny, 2017). The mid-Cretaceous Terrestrial Revolution occurred because splitting continents led to increasingly patchy, heterogeneous environments and isolated populations, both of which increase diversity over evolutionary time, while roughly simultaneously the radiation of angiosperms opened up new niche spaces for diverse lineages to occupy.

My aim in this Section is to examine philosophical accounts of narrative. Along the way, I'll focus on three questions. First, are narratives distinct from other kinds of explanation? Second, what are narratives for? That is, what kind of epistemic value does a narrative bring with it? Third, questions of realism: to what extent do historians discover narratives, and to what extent do they construct them? Realism in this context is understood in a particular way. Both realists and anti-realists agree on the past existing and on our having knowledge of it. They come apart on the nature of narrative events, that is, the aspects of the past that are highlighted as being significant in an historical narrative. For anti-realists, attributions of significance are the projections of historians, whereas for realists there is a mind-independent fact of the matter about historical significance; anti-realists think narrative structure is imposed on the world to make sense of it for us, whereas realists think narrative structure, in part, reflects how the world is.

Much discussion of narrative is rooted in Carl Hempel's account of scientific explanation, which emphasized the roles of laws: a scientific explanation cites some initial conditions, as well as some necessary laws, then shows how the thing-to-be-explained follows deductively from these (Hempel, 1942). Historians rarely cite anything approaching a necessary law, and their explanations don't appear to be deductive. Hempel's response was to consider narratives *explanation sketches*: proto-explanations. Others objected, arguing that history has a distinct mode of explanation, with a different logic and structure. There were differing views on how the logic differed. William Dray, for instance, argued that the split between historical and scientific explanation lies in the modal character of the information communicated (1957). Scientific explanations deductively infer how the event actually occurred; historical explanations demonstrate how the event could possibly have occurred (see also Gallie, 1964). The disagreement between Dray and Hempel, in my view, founders because of reliance on the contrast between how-actually explanations, which rely on necessary laws to tell us what in fact happened, and how-possibly explanations, which simply provide sufficient but unnecessary conditions. As we saw in the discussions of laws, above, historical processes generate a wide variety of regularities: 'necessity' and 'possibility' are far too coarse to accommodate the various modal scopes involved in historical explanation; we'll need a richer conceptual toolkit.

Some have understood narrative explanation by equating it with intentional explanation, what Elizabeth Anscombe called 'reason-giving' explanations (Anscombe, 1975). Anscombe contrasted reason-giving with causal explanations, the difference being that one cites mere causes whereas the other cites the means-ends reasoning of an agent. This thought, I suspect, underwrites the (to me, baffling) occasional assertion that narrative explanations are not causal. I doubt the value of equating narratives with intentional explanation. For one thing, there are plenty of explanations that have narrative features – even within history – that are not 'reason-giving' in Anscombe's sense, of which my discussion of the mid-Cretaceous Terrestrial Revolution is one. For another thing, given the diversity of causal explanations (see next Section), it is far from clear to me why the category of intentional explanation ought to be granted special, non-causal, status.

Arthur Danto argues that narrative explanations capture a particular kind of historical knowledge (Danto, 1962, 1985). The historical versions of *whats* and *whens* are 'chronologies': ordered lists of events. The historian's job is to weave chronologies into narratives. What is so special about that? For Danto, past events gain significance in virtue of their relationship with future events. The evolution of flowering plants in the Cretaceous had momentous consequences. It shaped many aspects of the modern world, particularly enabling the wide variety of pollinating insects whose mutualisms with angiosperms are such a distinctive part of global biota today. Only in retrospect, in recognizing that how things are now is a result, in part, of how things were then, can this significance be recognized. The historian takes a chronology and, using a 'narrative structure' which highlights points in the chronology, links them together to show how past events gave rise to later ones. Telling a 'history of angiosperms', for instance, involves highlighting the mid-Cretaceous terrestrial revolution, while de-emphasizing other events. Further, Danto argues that for such events to count *as events* in the first place, the historian's hand is necessary. Chronologies do not, properly speaking, have events in the mind-independent sense. This is a form of anti-realism: events are not 'in the world' independently of historians (see also Roth, 2017).

Another anti-realist is Louis Mink, who emphasizes the relationship between literary fiction and historical narratives: both use narrative structure as a 'cognitive device' (see also Ricoeur, 2010). Narratives make sense of historical episodes, thus generating human understanding. For Mink these narrative structures are not, as it were, in the world waiting to be discovered. Rather, we project them: the past is carved up into events according to the narrative structure we are using. One of his reasons for thinking this is that narratives do not aggregate. If there is a single narrative structure in the past then we should expect different narratives to be easily combinable. Yet they are not:

> A narrative must have a unity of its own; this is what is acknowledged in saying that it must have a beginning, middle, and an end. And the reason why two narratives cannot be merely additively combined ... is that [we need] a new unity, which replaces the independent coherence of each of its parts rather than uniting them (p. 197).

I find Mink's argument unconvincing. Explanations don't simply foreground and background, they also often distort and idealize: some features can be presented in exaggerated or simplified ways to aid in understanding, or emphasize their importance, or to demonstrate the affinities of that event with salient others. Because of these perfectly legitimate explanatory practices, mutually compatible explanations of the same event will not necessarily aggregate. But this doesn't lead to anti-realism about historical events, as the distortions are a feature of how the events are fitted into explanations, not of the events themselves.

David Hull is a realist. For him, where Hempel-style scientific explanations gain their unity from the laws appealed to, a narrative explanation is unified as a coherent whole, because it is about a particular kind of thing, a 'central subject' (Hull, 1975; Currie & Walsh, forthcoming). We might equate central subjects with token-historical-kinds. The history of a lineage, the evolution of *Triceratops*, say, is not coherent in virtue of there being some general law governing how they evolve. Instead there is an individual – a single topic – that the explanation tracks across time. And it is that single topic which grants the explanation coherency, and its narrative form. Hull is a realist because there is some object in the world that grounds narrative: we don't project the narrative onto the central subject, rather our narrative reflects features of the central subject.

John Beatty connects narrative explanation with contingency (Beatty, 2016, 2017). On Beatty's account, a narrative contrasts a particular sequence with close-by 'narratively possible' could-have-beens. Perhaps the angiosperm radiation was dependent on the break-up of Gondwana (let's say the continental spread was required to weaken the dinosaur-gymnosperm regime, which gave angiosperms their evolutionary opportunity). Under this scenario, the narrative highlights the contingency of the end-Cretaceous revolution on geographical changes. By tracing the branching moment (in this case, the continental break-up) we contrast how things actually went with how things could have gone in relevantly different circumstances.

> [Narratives] relate what happened, one event at a time. All narratives do that. But some … do more; they relate what happened, one event at a time, while indicating that some of those events might not have happened, but did. And they also indicate that what did happen, vs. what might have, was consequential (2016, p. 34).

Beatty adopts a loose, inclusive notion of 'narrative' and highlights particular narrative types as being particularly interesting. This is promising, and it is a strategy I'll follow below: instead of wrangling about just what a narrative is, we can ask what a narrative is good for. As I'll expand in Section 3.3, Beatty's answer is that narratives are particularly good for explaining change via contrast with other possibilities: 'tracing one path through a maze of alternative possibilities, and alluding to those possibilities along the way, is what narrative does particularly well' (2017, p. 34).

There are stronger and weaker ways of characterizing questions about narrative distinctness. Paul Roth puts things in a particularly strong way:

> [the distinctiveness of narratives turns on] whether or not an explanation in this form can also be non-narratively structured. That is, does it allow for paraphrase into some other, non-narrative explanatory form? For if so, then whatever explanatory import such a narrative seemingly possesses revealing how things at the beginning of a time series came to be what they later were turns out to be inessential for purposes of explanation (2017, p. 43).

The thought goes: I can capture scientific explanations using logical form (the scientist cites some initial conditions, cites some law-like regularities, and then derives the thing-to-be-explained from that), and this logical form explains something about why the explanation is an explanation (that is, it goes beyond mere description) because (1) it shows how the thing-to-be-explained had, or at least was likely, to happen given those initial conditions; (2) it unifies the thing-to-be-explained as an instance of a type of event. To defend narrative explanation, you must first show that I cannot translate narratives into scientific explanations. For if I can, then there is nothing distinctive about them. Then, to show that narratives in fact explain, you should convince me of their explanatory powers.

Building on Danto, Roth takes the answer to lie in 'narrative sentences'. In a narrative sentence, the thing-to-be-explained, and the things-doing-the-explaining are not independent in a way that allows explanation to take the form of a valid logical argument. In his words, they are 'indetachable'. Consider the term 'mid-Cretaceous Terrestrial Revolution'. This points to a stage in history (a point, or series of points, in a chronology) and (1) identifies it as being the middle stage of something (the Cretaceous), (2) implies some significance for it in light of future events. Consider this sentence: 'The mid-Cretaceous Terrestrial Revolution set the stage for the modern world'. This is what Danto calls a 'narrative sentence'. It is about some earlier time (the revolution), but is defined in light of its significance at later times. These 'refer to at least two time-separated events though they only *describe* and are only *about* the earliest event to which they refer' (Danto, 1962, p. 147, italics in original). So although the sentence is about events, say, 100 million years ago, its truth depends on events occurring much later. Roth argues that the links between the earlier and later states in a narrative explanation cannot be captured in a Hempel-style explanation because the thing-which-is-explained and the thing-which-does-the-explaining are inseparable. Because of this inseparability, any argument with one as the conclusion and the other as a premise would be circular. Importantly for Roth, this feature of narrative sentences is necessarily retrospective: it is only in light of taking an historical perspective, looking back, that the significance of past events can be identified. Roth takes this to mean that historical events are constructions – this is another route to anti-realism about historical events.

Roth and Danto, then, adopt a kind of philosophical challenge for the legitimacy of narratives: identify some feature in virtue of which narratives are logically different from – that is, not intertranslatable with – other forms of explanation, particularly Hempel-style explanations. In Section 1.1, I worried about these kinds of philosophical challenges: they often leave accounts beholden to fundamental philosophical analysis and they can be dangerously detached from the phenomenon we're interested in explaining. There are two further reasons to reject Roth and Danto's approach. One has to do with the diversity of accounts of explanation on the market: philosophers of science increasingly recognize a plurality of explanatory-types, and in light of this, demanding that each type be logically different to each other (particularly Hempel's account), is insufficiently sensitive to the more pragmatic, pluralistic turn the philosophy of explanation has taken. As I'll expand on this argument in the next subsection, I'll spend more time here on another reason: the connection to anti-realism.

Because Roth and Danto's challenge links narrative to linguistic constructions it is closely aligned to anti-realism. As we've seen, anti-realism is, at base, the idea that narratives (or events in a narrative) are 'projected onto' the world: narrative structure is not part of nature. A realist about narrative will argue that at least some aspects of the world's structure are captured by narratives. To be clear: both realists and anti-realists agree that narratives are legitimate and important knowledge-generating tools. The question turns on whether these tools are, as it were, purely cognitive; only working to make sense of an otherwise senseless world, or whether they additionally track real features that the world has.

Anti-realism cannot do justice to a class of substantive debates common in historical science. Roth has argued that although narratives are underdetermined by evidence, we can judge between narratives in terms of 'their fruitfulness in guiding research and their resources for solving problems' (Roth, 1988, p. 12). For anti-realists, then, debates about history should turn on either (1) whether or not such-and-such events occurred when they did (that is, disagreements about chronologies) or (2) whether or not such-and-such events are significant (disagreements about narrative structure). This latter kind of debate turns on facts about us as knowers (problem-solving or fruitfulness, say) not on the state of the world. However, historical scientists commonly debate the significance of historical events in empirical terms. Consider models of dinosaur extinction. One hypothesis claims the impact event knocked out a thriving biota, and so was highly significant, whereas other hypotheses highlight previous events (the mid-Cretaceous revolution, for instance) that critically weakened the biota, thus lessening the impact's significance. Such debates are neither simply about what events occurred in what order, nor about which way of framing the events is the most psychologically satisfying or fruitful. They are about whether some narrative or the other is true (or, if you want, truer). Scientists engaged in these debates, as we've seen, marshal empirical information in determining significance: whether the events of the mid-Cretaceous could have weakened dinosaur ecosystems as suggested, whether the patterns of disappearances from the fossil record match the hypotheses, and so on. No doubt historical disputes are subtle matters (Currie & Walsh, 2019), but they are substantive, empirical and, unless anti-realists can show how to satisfactorily accommodate them, this motivates realism about historical events. There is often a fact of the matter about whether an event is significant.

If we are to reject the need for the philosophical bar Roth has set, how should we approach narratives in science instead? Recall John Beatty's earlier suggestion (2016) that instead of asking what narratives are, we ask what they are good for. I take this to be in line with the practice-based approach we have followed. A practice-based approach to narrative, I take it, involves (1) examining narratives as they appear in science, and (2) considering why narratives might be useful for scientists' purposes. This doesn't require showing that narratives differ logically from other kinds of explanations, but rather requires identifying features of narratives that make them suitable for various kinds of task. To do this, I want to first say something general about explanation, and use this to bring out the relevant narrative features.

3.2 Explaining as Situating

Philosophers increasingly recognize a variety of explanatory forms. A common theme of the last 30 years has been *ecumenism*: the idea that there are different kinds of legitimate explanations (more controversial) and that, for any one phenomenon, there are at least two legitimate explanations applicable (less controversial) (Jackson & Pettit, 1992; Sterelny, 1996). We've seen that discussion of narrative often leans on an explicit comparison with Hempel-style, covering-law explanations. It is taken that explicating something in-principally different from law-based explanation is required to legitimize narrative explanation. But philosophers of science have described a plethora of ways of shifting from whats and whens to whys and hows. And this undermines demanding a logical distinction between explanatory forms. As I'll explain, I think of these different explanatory forms as different ways of 'situating' a target. Let's consider a few examples, before thinking about narratives more carefully. My discussion of non-narrative explanation will be purposefully sketchy, as my purpose is simply to give a flavour of some options.

As we've seen, 'deductive nomological' or 'covering-law' explanations situate a target in terms of a necessary law and a set of initial conditions: they show how, given a set of dynamics and an initial condition, the outcome simply had to happen. The target is situated in an argument with a deductive structure. This often leads to its being considered in abstract ways. This abstraction often makes covering-law explanations useful for considering the target as a token of a type.

Alternatively, we might situate a phenomenon mechanistically (Craver, 2007; Machamer et al., 2000). A mechanistic explanation accounts for some phenomenon by identifying the entity that produces it. The entity is decomposed to its constituent parts, and the entity's capacities are explained by considering the causal powers of those parts and their relations. Investigation of the function of the thickened skulls of *Pachycephalosaurus* look mechanistic in this way. The skull is considered in terms of how tightly packed and dense are the materials that make it up, and this is used to try to understand how much stress it could undergo. The causal power of the skull – its resilience – is understood in terms of its constituents and their activities.

Another way of situating a phenomenon is via *minimal modelling* (Weisberg, 2007). A minimal model is an attempt to isolate the 'essential dynamics' of some system and use these dynamics to explain the system's behaviour. As Michael Weisberg put it, 'the key to explanation is a special set of explanatorily privileged causal factors. Minimalist idealization is what isolates these causes and thus plays a crucial role for explanation' (2007, p. 103). The earlier appeal to weakened terrestrial dinosaur ecosystems in explaining their extinction looks like this. The ecology is represented using a simplified model, which is taken to capture what matters concerning the resilience of that kind of ecosystem. Minimal modelling involves, then, focusing on a few features of the target and considering their similarities to an idealized version of those features.

Explanations, on the pluralist view I'm pushing, take non-explanatory information (descriptions) and situate this in a way which makes sense to us, which has some explanatory pay-off. The difference between what and why is that whys require situating the what in a way that generates understanding. The wide variety of explanatory strategies reflects the wide variety of ways in which we might situate our target, and the variety of explanatory pay-offs (kinds of understanding) we might desire concerning our target. My argument for history mattering for knowledge is based on the thought that some targets – peculiar ones – demand narrative explanation. Before getting to that argument, we should return to narratives: how do they situate?

As we've seen, Danto thinks about narrative in terms of 'temporal structures'. Temporal structures highlight some aspects and background other aspects of a chronology, linking them together and creating a narrative. Similarly, Hull highlights the central subjects around which the coherency of the narrative is woven. Narratives are token, multi-step trajectories. The target is temporally situated, taken as a step along a particular process, or the culmination of that process. But, following Beatty, narratives do not simply situate temporally ('relating what happened one event at a time') but also modally. In foregrounding particular events and processes as being significant, their causal powers – how they shape the trajectory – are emphasized. Including the rise of angiosperms in our narrative of dinosaur extinction links these two events: the occurrence or otherwise of the radiation made an important difference to the dinosaur extinction. Narratives situate a temporal trajectory within a possibility space.

Because narratives take the form of a trajectory linked together through varying degrees of significance, they can accommodate a lot of *complexity*. And not simply complexity in the sense of there being many interacting parts: historical explanations often demand different components operating at different scales. The mid-Cretaceous Revolution is like this: explaining it requires reference to geographical changes, the relationship between geography and biodiversity, specific information about the developmental plasticity of angiosperms, why dinosaurs were less able to adapt to the new environments and so forth. Narratives have an expansiveness and the capacity to accommodate a variety of features that makes them well-suited to these kinds of complexity.

Another feature narratives are adept at capturing is *disunity* or *uniqueness*. There's reason to believe dinosaurs have distinctively plastic morphological development, and this might explain their apparently inflated biodiversity in the late-Cretaceous. In understanding this development, we typically want to know what makes dinosaurs different from, say, crocodilians or birds (or even mammals). Because narratives focus on token trajectories, they are well placed to capture disunity. Finally – and perhaps most obviously – narratives capture *temporality*. The mid-Cretaceous Revolution wasn't simply a period of change or disruption, it involved a shift from one business-as-usual to a new business-as-usual. Often such shifts are path-dependent, their occurrence turns on the ordering of events, and such shifts can involve changes in dynamics, illustrated by the decrease in the resilience of dinosaur communities in the late-Cretaceous. That narratives represent by picking out events from within a chronology makes them well-suited for capturing this temporality.

Narratives, then, are good for situating explanatory targets which are complex, disunified and temporal. It doesn't follow that other explanatory forms cannot also manage to accommodate these features: I've argued that we shouldn't accept Danto and Roth's challenge, instead considering explanatory forms in terms of what they are good for. Along these lines, I also don't think that narratives capture complexity, disunity and temporality by definition (I've argued elsewhere that narratives are sometimes simple, Currie, 2014, and Glennan, 2010 argued that mechanistic explanation can also accommodate some of these features), rather, these are properties well-suited to a narrative treatment. So, rather than seeing the variety of explanatory forms as being in competition, we can view them as complementary: they are different explanatory strategies; they are different 'ways of situating'. This stance is easily adopted if we refuse to provide logical, absolute and fundamental grounds for separating explanatory forms. Whether or not ultimately I could translate a mechanistic explanation into a narrative or minimal model explanation (or vice versa), if we think of them as explanatory approaches or strategies, they are nonetheless different ways of explaining in practice. Note also there is no tension between me characterizing narratives as I have while refusing to provide a strict, logical definition of them: my characterization serves to pick out a set of explanatory forms, without asking whether they are translatable into others.

I do think there is a way of unifying these disparate accounts, however. They are all ways of shifting from descriptions to explanations via *situating*. A mechanistic explanation situates the target in terms of the activities and causal relations between component parts; a minimal model explanation situates a target as imperfectly instantiating a set of causal dynamics; a covering-law explanation situates a target as being the necessary outcome of a set of initial conditions; a narrative explanation situates a target as being part of, or the culmination of, a causal trajectory over time.

Although it is pretty shallow, this 'situating' account of explanation does important work for me here: it allows us to re-tool Roth and Danto's challenge for narrative explanation in weaker, more pragmatic terms. Specifically, are there targets which, as it were, *demand* that they be situated in narratives, those which are particularly well-suited to narrative treatments? I think there are: targets exhibiting peculiarity.

3.3 Peculiar Explanations

I've introduced the notion of narrative explanations, explanations that account for particular events, processes or entities by situating them within a multiple-part trajectory and contrasting them with other possibilities. In Section 2.3 I argued that distinctively historical processes lead to peculiarity: often generating their own dynamics, or kinds, via path-dependent, transformative sequences. Here,

I want to bring these aspects – narrative explanation and peculiarity – together, and argue that narrative explanations are particularly well-suited to, perhaps even necessitated by, historical peculiarity. Narratives are good at accommodating – situating – peculiarity. And it is the historian's interest in, and concern for, the peculiar that explains their penchant for narrative.

This view has most in common with that of John Beatty. Recently, Beatty has provided an account of narrative-worthiness (what narratives are good for) in terms of their capturing 'narratively open' possibilities (2017) and, so far as I can tell, what I have to say here coheres with his account. But he has also provided (2016) an argument rooted in the notion of 'choice-points' or 'eventful-events', and where his and my accounts come apart in this regard is instructive. Beatty emphasizes how narratives highlight moments where some trajectory could have gone differently, and from this argues that narratives are worth telling when they capture 'branching events', moments in an actual history where things could have gone one way but instead went another. If the dinosaur extinction was indeed caused by a big rock hitting the Earth, then that impact event was a choice-point: history's path switched from one trajectory and careened into another.

While both Beatty and I emphasize the modal character of narratives, he conceives of this modal character in terms of a tree with distinct branching moments. A moment in a narrative is a choice-point depending on 'whether it leaves open or forecloses the possibility of reaching that goal' (p. 36): choice-points are when the contingent becomes locked in, or (perhaps, Beatty isn't explicit about this), possibilities are opened up. If, say, the asteroid impact 65.5 million years ago was necessary for the dinosaur extinction, then it is a choice-point: if the rock hadn't hit, then the dinosaurs wouldn't have gone extinct. Choice-points are a kind of possibility bottleneck. I'm going to sketch three arguments against the idea that capturing choice-points is what narratives are good for, which will set us up for my positive account. In fairness, Beatty might be making a claim about sufficiency rather than necessity: choice-points might simply be one of many things narratives are good for. And indeed, my account doesn't exclude choice-points: choice-points are one common feature of peculiar histories, but they are neither necessary nor sufficient for them.

First, Beatty (like myself) is plausibly read as a realist: narrative events are not things we project onto the past, but things we discover in the process of investigating it. I worry that Beatty's account is in danger of conflating real stories with good stories. No doubt good stories emphasize change and contingency: a story where nothing changes, or very little is at stake in the changes that do happen, is a boring story. But there is nothing to guarantee the world follows the dictates of good storytelling. Real stories might often be boring. If narratives are the modus operandi of historical science, and narratives are only worth telling when things are 'interesting' (that is, when there are 'choice-points'), then either large amounts of history will go unexamined or we might begin over-projecting contingency onto the world. This challenges realism by introducing systematic bias into historical explanation.

Second, and again assuming that narratives are supposed to be a privileged mode of explanation in history, Beatty's view prioritizes explanations of change over stability. The world is often a chaotic, complex place: we might think that what is remarkable is not that things change, but that sometimes they don't. Consider punctuated equilibrium models of speciation. As we saw in Section 1, these are contrasted with gradual phyletic speciation. According to punctuated equilibria, a species' life is marked by stability which then collapses (Turner, forthcoming). Derek Turner (2018) has argued that one of the central reasons for paleontology's uneasy relationship with more traditional Darwinian approaches to biology is (in essence) the former's focus on explaining change rather than stasis. For the biologist focused on explaining change, stasis in the fossil record is mostly ignored, but for paleobiologists (particularly those interested in punctuated models of speciation) stasis is a phenomenon: it is something to be explained, and, I think, explained in a narrative way. John Dupre and Dan Nicholson make this point much more generally:

> In any scientific enquiry it is necessary to distinguish what requires explanation from what is background, taken for granted. The orthodox substantialist position of modern science typically takes this background to involve stability: if nothing changes, then nothing requires explanation . . . For a process, however, change is the norm, and it is relative stability that takes priority in the explanatory order (Dupre & Nicholson, 2018, p. 14).

This leaves open discussions about when stasis deserves explanation and whether Beatty's appeal to 'eventful events' could be adapted to stasis. I'm not sure: we're sometimes interested in stability in terms of its origins – in virtue of what did the angiosperm-insect alliance arise and what explains its subsequent stasis? The earlier conjunct plausibly involves an eventful-event, but the later does not. We're also interested in stasis when we expect instability: given environmental variability, we might be surprised as to the long stability some lineages appear to exhibit. Perhaps we could consider the events as eventful in the sense that other possibilities were available but not taken. However, I think some narratives point to the opposite of eventful-events: we explain why various different paths weren't open in the first place, why the events weren't eventful. 'Leaving open or foreclosing' a possibility is not what makes stasis worthy of narrative: it is sometimes the continuity that is surprising and deserving of explanation.

Third, eventful-events and choice-points are evocative of possibility bottlenecks, momentary, discrete events where things could have gone one way or another. But change needn't be sudden, and possibilities needn't switch one way or the other at particular times and places. As we saw in Scannella et al.'s (2014) discussion of *Triceratops*, sometimes speciation is a gradual, incremental business. If *Triceratops* evolved phyletically, then there may have been no moment where things went one way or another. That is, there may not be a clearly delineated event or set of events which determined the trajectory. In such circumstances there are no eventful-events and thus nothing worth narrating on Beatty's view. Focusing on choice-points misses both incremental and punctuated models of speciation. They miss incremental speciation because speciation in such cases are trends, rather than eventful-events; they miss punctuated speciation because in such cases the explanatory focus is on stasis. Why should the tumultuous change from the dinosaur-gymnosperm Mesozoic to the insect-angiosperm Cenozoic take explanatory precedence over the remarkable stability of those two systems? I see no good reason to think it should.

These three objections have parallels for someone emphasizing stability and gradualism at the expense of fragility. History contains multitudes of kinds of change and stability and we shouldn't privilege one over the other without good reason. Although Beatty is right to think that alluding to possibilities and relating them to the actual in a step-by-step manner is what narratives do well, tying this to choice-points focuses too much on change against stability, and points against trends. Peculiarity, I think, includes Beatty's insights but goes further. History, I think, matters for knowledge because peculiarity demands narrative explanation. Recall our second question from the introduction: 'Does something's history matter to the knowledge we can have of it?'. In answering this question, I want to identify a feature of history and demonstrate that some forms of knowledge are well-suited to capturing it. Peculiarity is not necessarily a feature of every past event: some things may not be peculiar, or may be only minimally so. Peculiarity, I hope, identifies the aspects of the past that are truly historical (that involve, for instance, Collingwood's new forms). How do we show that some forms of knowledge are better suited to history (that is, peculiarity)? Recall Roth's challenge: the distinctiveness of explanatory form turns on intertranslatability; if I can reword a narrative explanation into a covering-law explanation then there is nothing special about narratives. By the contrasting approach I prefer, instead of asking in-principle questions, we consider how different explanatory strategies generate understanding via different kinds of situating. Our challenge, then, is to show that some forms of knowledge (narrative explanations) are particularly good (or, stronger: uniquely good) at situating peculiarity.

Recall Franklin-Hall's distinction between synchronic and diachronic kinds. The former carves up the world in terms of essential properties, or at least via properties held at a time-slice. The latter carves the world either in terms of process-types (type-historical kinds) or token trajectories (token-historical kinds). The mid-Cretaceous Terrestrial Revolution is a token-historical kind. It is an event drawn out over millions of years, sweeping up the biodiversity of a wide range of taxa, from insects to mammals and birds. As we've seen, explaining the revolution requires:

(1) complexity: multiple different components, operating at different scales (from the relationship between continental break-up and radiations, to ecological resilience in different communities, to the browsing habits of dinosaurs versus other taxa);

(2) disunity: although the revolution is in some ways similar to other radiations, it is unique, the details that made a difference are particular to that trajectory (the flexibility of angiosperms, or dinosaurs' unsuitability to consuming them, for instance);

(3) a temporal dimension: angiosperms went from their first emergence to near-dominance by the middle of the Cretaceous, understanding this requires a sequence lain out over time.

These three are just the features I discussed concerning narrative explanations. Such explanations are tailor-made for situating complex, unique and temporally extended sequences. The peculiarity of the Cretaceous events we've considered, the robust angiosperm-insect alliance, the newly fragile dinosaur ecosystems, the radiating mammals and birds, and so forth, demand an explanation that can accommodate their complexity, interconnectedness and long temporal scale. This calls for being situated in a token, multi-step (and multi-scale) trajectory. That is: a narrative. History matters because its peculiarity demands a narrative treatment; narratives are required to accommodate peculiar targets. History matters for ('gets into') a peculiar target because the target's modal profile is specifically a result of the processes that produce and maintain them; if we want to explain why our target changed as it did, or remained as it was, we must appeal to features of those processes.

This view has at least two upshots.

First, peculiarity accommodates – and potentially supports – realism about narrative events. Recall that both realists and anti-realists believe that there are facts about the past; however, anti-realists think the narrative events – the structure their explanations lay onto the world – are projections, cognitive tools, not representations of how things were. Realists agree that narratives are cognitive tools for generating understanding, but go further to say that successful narratives in fact reflect the structure of the world. Peculiarity is a real feature of the past: diversity-boosting processes really do lead to new stabilities, fragilities and novelties over time. And historical narratives represent that peculiarity. It may be that the significance of some events is only recognizable in terms of the peculiarity they generate downstream. Say, Gondwana's breaking apart is important because of the subsequent terrestrial revolution. But it in no way follows from this that we construct those events, or that their existence depends on historians recognizing them. Its significance for the mid-Cretaceous revolution is not a projection, but real. The Cretaceous' peculiarity is something we discover and that we can have (and paleontologists do have) substantive disagreements about.

Second, as we saw, peculiarity explains the 'localness' of scientific knowledge. In Section 1, we saw how evidence-generation in historical science often depends on highly idiosyncratic knowledge tailored to the conditions at hand. In Section 2, we saw how it relied on exception-ridden, context-sensitive (yet nonaccidental) regularities. Given the peculiarity of history, this is just what we should expect: diversity-boosting processes generate both fragility and robustness in particular (peculiar!) processes. And these are only capturable in local ways (see also Beatty, 1997). The fragmented nature of historical knowledge is a response to the fragmented nature of history. This is why general, all-purpose accounts of evidence or explanation are of such limited use in understanding knowledge-generating practices. To be clear, none of this suggests that merely being in the past is what makes something epistemically local: our current world is without doubt peculiar, as will the future be. But it is in being peculiar that something's history matters.

Concluding Discussion: What Are Sciences of the Deep Past About?

I've argued that history matters for knowledge. First, the past always features in evidential reasoning, because the provenance of data always matters for its deployment as evidence. Second, history is peculiar, and peculiarity demands narrative explanations. Neither of these big-picture points are restricted to knowledge of the deep past (shallow pasts are included, and indeed we've seen how history matters to knowledge of current and future events too!), but both tell us something important about what our knowledge of the deep past is like. Like all empirical knowledge, the provenance and journeys of data – their past – makes a critical difference. Further, our knowledge of the deep past is likely to take a complex, narrative form. Our understanding it depends on it being situated within a trajectory moving through a space of possibilities. Getting that knowledge depends on a complex social practice of iterative investigations involving theoretically grounded expectations, field-work, data-processing and storing, and data deployment in analyses and evidential claims. In the Introduction, I also discussed one clear way in which knowledge of the deep past matters: some important questions are only answerable from that perspective. Evidence from the deep past is required to inform theorizing about, for instance, the origin of the universe, the formation of the Earth, the shape of life and the unfolding of human culture across the globe.

But what is the aim of sciences concerned with the deep past, what is the research ultimately about? This is my final question. Attending to philosophers of historical science (myself included), a kind of answer emerges: historical science is about telling the history of their subjects, tracing their unfolding and their eventual conversion into traces. Carol Cleland is perhaps the most explicit on this score: 'Hypotheses concerning long-past, token events are typically evaluated in terms of their capacities to explain puzzling associations among traces discovered through fieldwork' (Cleland, 2011, p. 552).

The point of historical science is to explain the traces we see now. Are, for instance, the patterns in *Pachycephalosaurus* skull fossils, (some domed, some flat) best explained in terms of their having separate evolutionary histories (that is, being different species) or in terms of being part of a shared ontogenetic sequence? Answering these questions uncovers particular histories. Aviezer Tucker also often talks in such terms:

> The historical sciences are concerned with inferring common causes or origins: contemporary phylogeny and evolutionary biology infer the origins of species from homologies, DNA, and fossils; Comparative Historical Linguistics infers the origins of languages from information preserving aspects of existing languages and theories about the mutation and preserva-tion of languages over time (Tucker, 2011, p. 20).

Like Cleland, Tucker takes the point of historical science to be provisioning particular histories, revealing the unfolding of whatever their subject is. No doubt this is an important goal: for instance, knowledge of such unfolding can form the basis of our understanding our own time and future (Currie, 2018a, chapter 12). But do such narrow conceptions of historical science really capture what the study of the deep past is about? That is, do they exhaust the point of studying the deep past? I don't think so, and I think our discussion thus far has shown why.

In contrast to philosophers who examine historical reconstruction from the perspective of methodology and epistemic puzzles, archaeological and historiographical theorists sometimes take a broader perspective. In archaeology, for instance, there is a tradition of contrasting accounts that emphasize archaeology as being about traces and trace-based reasoning and those that take archaeology as being about . . . something deeper. John Barrett captures this contrast particularly eloquently:

> Most outside observers, along with all too many practitioners, define archaeology in the banal terms of digging, discovery of old things, and the physical analysis of those things. It is from this perspective that the history of archaeology is written as the development of techniques of recovery and material analysis. This consigns archaeology to the role of antiquarianism . . . such a negative perception surely contrasts with the more challenging view that archaeology could offer itself, namely as an enquiry into the full chronology and global extent of humanity's place in history (Barrett, 2016, pp. 133–134, references removed).

What is it to enquire into 'the full chronology and global extent of humanity's place in history'? One contrast might be between cataloguing material remains and inferring from them to the past, but I think more is going on. Barrett emphasizes the diversity of human life-ways, and the role of material conditions in enabling, shaping – but not determining – them. A culture's material conditions do not in themselves decide everything about that culture, but they provide the stage upon which that culture acts. If storing information in written form is required for certain kinds of enquiry, then these enquiries will be difficult for nomadic groups who can't cart tons of books around. Archaeology, for Barrett, is about getting a grip on how material conditions enable human life-ways: 'An engagement with an archaeological record often prompts the need to *explain* its formation. An engagement with the material conditions of human possibility on the other hand prompts the desire to *understand* those conditions' (p. 137).

Relatedly, Joan Gero has argued against a focus on certainty in archaeology and in favour of preserving the ambiguity of archaeological evidence and theorizing: roughly speaking, archaeologists should wear their interpretation's underdetermination on their sleeves. Part of her argument is that this underwrites a more discursive and reflective science: '[archaeologists should] work towards an archaeology that *interrogates* the past instead of advancing conclusions as exclusively and exhaustively final and "right"' (Gero, 2007, p. 313).

On one picture, archaeology is in the business of uncovering, explaining and understanding the actual history of the past; on another picture, the science is in the business of interrogating the past to understand the conditions of human existence. These richer views of the point of history emphasize (1) understanding the conditions of possibility, and (2) a two-way, discursive approach to knowledge generation. The narrower accounts philosophers imply emphasize (1) explaining traces in terms of past events, and (2) telling histories. These two accounts need not be mutually exclusive, indeed Gero argues that practices denying ambiguity 'contradict the long-term archaeological interests of accumulating accurate information about the past' (p. 313).

You might (perhaps echoing Collingwood) think that archaeology carries this value because its subjects are human, intentional subjects. I don't think so. I'm going to close by arguing that the features folks like Bennet and Gero highlight for human history are not exclusive to humanity: they are values of investigations of the past generally. Sciences of the deep past are about uncovering the material conditions of existence, and are two-way discursive practices. This is seen when we connect these discussions with two features from earlier: the iterativity and dependency between idiographic and nomothetic investigations, and the peculiarity of history.

In Section 2.1, I argued that in practice historical science is ultimately neither nomothetic (concerned with understanding regularities) nor idiographic (concerned with particular histories) because both kinds of knowledge are mutually dependent. To understand whether or not *Triceratops* and *Torosaurus* are the same species, we need to appeal to general understandings of biodiversity patterns, as well as our expectations about ontogenetic stages in those critters: the idiographic leans on the nomothetic. And those nomothetic theories are themselves built from interactions with particular cases. I highlighted the iterative nature of this interaction. Hasok Chang, analysing the history of measurement, similarly emphasizes what he calls 'epistemic iterativity': 'What we have is a process in which we throw very imperfect ingredients together and manufacture something just a bit less imperfect' (Chang, 2004, p. 226).

As Kevin Elliot pointed out (2012), such iterativity occurs both at the level of knowledge claims – between expectations about biodiversity and the particular taxonomic affiliation of specimens, say – and at the level of methods: between techniques of fossil preparation and their deployment as evidence about the past. This iterativity looks, at least to me, like exactly what we should expect from an interrogation of the past, of 'conversing' with other times. Iterative back-and-forths also play a critical role in Bob Chapman and Alison Wylie's explanation of the power of trace-based reasoning (my own notion of 'methodological omnivory' (2015, 2018) draws from the same well). Generally speaking, I think, science often progresses via the iterative construction and destruction of scaffolds (Walsh, forthcoming), and this requires a two-way iterative process. This, I think, is a pretty close match to discussion of 'discursive' investigations of the past. It may be that I've missed some kind of normative dimension: our targets' lacking intention could mean we don't judge, criticize or empathize. But we can certainly compare. Our current world, angiosperm-rich with divided continents, differs profoundly from that of the early-Cretaceous. And these comparisons may carry crucial lessons: as our world becomes 'smaller' through global connections, perhaps in some ways we return to Gondwana's relative homogeneity, and the diversity-dampening that implies. Moreover, seeing the fall of the dinosaur-gymnosperm alliance reminds us that long term changes can undermine the most apparently stable of systems. Our increasing capacity to intervene at global levels should lead us to worry about the stability of the global systems we depend on. Dinosaurs managed millions of years, we've barely gone a hundred thousand. The discursive nature of investigations of the deep past leads to reflection on our current state and our future states, then: iterativity is not limited to knowledge and method.

History's peculiarity not only motivates narrative explanations, but also underlies concern about the conditions of existence. Many historical enquiries are not simply about telling the Earth's history, but about understanding the ways in which Earth could have been (and perhaps could be). A world with a biology-produced, oxygen-rich atmosphere is a different prospect from one with depleted oxygen. A world with flowering plants is, in many ways, radically different from one without them. A world with domesticated plants, agriculture, offers different opportunities – and costs – from one without them. Understanding peculiarity requires more than tracing a particular history, but getting to grips with how previous conditions laid the groundwork for, enabled, dampened and triggered later conditions.

In investigating the deep past, then, we do not simply uncover a sequence of narratives. We also learn about the wild variety and diversity of history, and the modal properties underlying and enabling those forms. We don't simply learn about how continents breaking apart and flowering plants evolving led to a dramatic revolution in many of the world's biotas, we also learn about the relationships between biodiversity, geography and adaptation. We learn about the conditions required for the emergence of stable dynamics like the insect-angiosperm alliance and the conditions leading to the dissolution of stability, as with the dinosaur extinction. We don't simply tell a history, but uncover the conditions of existence. Rich views of the purpose and nature of history are not restricted to human history, therefore. The deep past carries this value.

And this carries with it a lesson for how philosophy of science should be done. In the Introduction, I presented a methodological dilemma: how can my analysis of historical knowledge be normative if it is based on descriptions of practice? The answer is that the 'descriptive/normative' dichotomy is too coarse to capture the nature of the argument being made. Rather – and similarly to history – in examining scientific practice I bring my preconceptions about science (the nature of evidence, explanations and so forth) and those preconceptions shape how I go about that examination. But scientific practice will not bend any way I desire: it pushes back and in turn shapes those conceptions. That is, the examination is iterative, mingling normative philosophical theory with descriptions of scientific practice. There is no methodological dilemma here.

And so, history matters for knowledge in many ways, and as we've seen this should lead us to de-emphasize synchronic characterization – avoid idealizing away from history – and adopt more local approaches to understanding and explanation. History doesn't just matter for scientific knowledge, then, it also matters for philosophy.

Bibliography

Anscombe, G. E. M. (1975). *Intention.* Cambridge, MA: Harvard University Press.

Armstrong, D. M. (1983). *What is a Law of Nature?* Cambridge: Cambridge University Press.

Barrett, J. C. (2016). Archaeology after interpretation. Returning humanity to archaeological theory. *Archaeological Dialogues* 23(2), 133–7.

Beatty, J. (2017). Narrative possibility and narrative explanation. *Studies in History and Philosophy of Science Part A* 62, 31–41.

Beatty, J. (2016). What are narratives good for? *Studies in History and Philosophy of Science Part C: Studies in History and Philosophy of Biological and Biomedical Sciences* 58, 33–40.

Beatty, J. (2006). Replaying Life's Tape. *The Journal of Philosophy* 103 (7):336.

Beatty, J. (1997). Why do biologists argue like they do? *Philosophy of Science* 64, S432–43.

Beatty, J. (1994). Theoretical Pluralism in Biology, Including Systematics. In L. Grande & O. Rieppel (eds), *Interpreting the Hierarchy of Nature: From Systematic Patterns to Evolutionary Process Theories.* San Diego, CA: Academic Press, pp.33–60.

Bell, M. (2015). Experimental archaeology at the crossroads: a contribution to interpretation or evidence of 'xeroxing'? In R. Chapman & A. Wylie (eds), *Material Evidence.* New York: Routledge, pp. 42–58.

Benton, M. J. (2010). The origins of modern biodiversity on land. *Philosophical Transactions of the Royal Society of London B: Biological Sciences* 365(1558), 3667–79.

Binford, L (1977). General Introduction. In L Binford (ed.), *For Theory Building in Archaeology.* New York: Academic Press.

Bokulich, A. (2018). Using models to correct data: paleodiversity and the fossil record. *Synthese,* https://link.springer.com/article/10.1007/s11229-018-1820-x#citeas

Bonnin, T. (2019). Evidential reasoning in historical sciences: applying Toulmin schemes to the case of Archezoa. *Biology & Philosophy* 34(30), 1–21.

Camardi, G. (1999). Charles Lyell and the uniformity principle. *Biology and Philosophy* 14(4), 537–60.

Chang, H. (2004). *Inventing temperature: Measurement and scientific progress*. Oxford: Oxford University Press.

Chapman, R., & Wylie, A. (2016). *Evidential reasoning in archaeology*. London: Bloomsbury Publishing.

Cleland, C.E. (2013). Common cause explanation and the search for a smoking gun. In V. Baker (ed.), *125th Anniversary Volume of the Geological Society of America: Rethinking the Fabric of Geology*, Special Paper 502 (2013), pp. 1–9.

Cleland, C. E. (2011). Prediction and explanation in historical natural science. *The British Journal for the Philosophy of Science* 62, 551–82.

Cleland, C. E. (2002). Methodological and epistemic differences between historical science and experimental science. *Philosophy of Science* 69 (3), 447–51.

Collingwood, R. G. (1976/1936). *Human nature and human history*. London: Ardent Media.

Colyvan, M. (2015). Indispensability Arguments in the Philosophy of Mathematics. In E. N. Zalta (ed.). *Stanford Encyclopedia of Philosophy*, https://plato.stanford.edu/archives/spr2019/entries/mathphil-indis/

Craver, C. F. (2007). *Explaining the brain: Mechanisms and the mosaic unity of neuroscience*. Oxford: Oxford University Press.

Currie, A. (forthcoming). Bottled Understanding: the role of lab-work in ecology. British Journal for the Philosophy of Science.

Currie, A. (2018a). *Rock, Bone, and Ruin: An Optimist's Guide to the Historical Sciences*. Cambridge, MA: MIT Press.

Currie, A. (2018b). The argument from surprise. *Canadian Journal of Philosophy* 48(5), 639–61.

Currie, A. (2015). Philosophy of Science and the Curse of the Case Study. *In The Palgrave Handbook of Philosophical Methods*. London: Palgrave Macmillan, pp. 553–72.

Currie, A. (2015). Marsupial lions and methodological omnivory: function, success and reconstruction in paleobiology. *Biology & Philosophy* 30(2), 187–209.

Currie, A. M. (2014). Narratives, mechanisms and progress in historical science. *Synthese* 191(6), 1163–83.

Currie, A. & Killin A. (2019). From things to thinking: Cognitive archaeology. *Mind & Language* 34(2), 263-79.

Currie, A., & Levy, A. (forthcoming). Why Experiments Matter. Inquiry.

Currie, A., & Sterelny, K. (2017). In defence of story-telling. *Studies in History and Philosophy of Science Part A* 62, 14–21.

Currie, A & Walsh, K. (forthcoming). Frameworks for Historians and Philosophers. HOPOS.

Danto, A. C. (1985). *Narration and knowledge*. New York: Colombia University Press.

Danto, A. C. (1962). Narrative sentences. *History and Theory* 2(2), 146–79.

Desjardins, E. (2011). Historicity and experimental evolution. *Biology and Philosophy* 26: 339–64.

Dilcher D. Towards a new synthesis: major evolutionary trends in the angiosperm fossil record.*Proceedings of the National Academy of Sciences USA* 2000;97:7030–6.

Dray, W. (1957). *Laws and explanation in history*. Oxford: Oxford University Press.

Dupré, J., & Nicholson, D. (2018). *A manifesto for a processual philosophy of biology. Everything flows: towards a processual philosophy of biology.* Oxford: Oxford University Press.

Elliott, K. C. (2012). Epistemic and methodological iteration in scientific research. *Studies in History and Philosophy of Science Part A* 43(2), 376–82.

Ereshefsky, M. (2014). Species, historicity, and path dependency. *Philosophy of Science* 81(5), 714–26.

Franklin-Hall, (in prep). Why are some kinds historical and others not?

Franklin-Hall, L. R. (2005), Exploratory Experiments. *Philosophy of Science* 72, 888–99.

Gaines, R. R., Briggs, D. E., & Yuanlong, Z. (2008). Cambrian Burgess Shale–type deposits share a common mode of fossilization. *Geology*, 36(10), 755-8.

Gallie, W. B. (1964). Philosophy and the historical understanding. New York: Schocken Books.

Gero, J. M. (2007). Honoring ambiguity/problematizing certitude. *Journal of Archaeological Method and Theory* 14(3), 311–27.

Ghosh, P., Bhattacharya, S. K., Sahni, A., Kar, R. K., Mohabey, D. M. & Ambwani, K. (2003). Dinosaur coprolites from the Late Cretaceous (Maastrichtian) Lameta Formation of India: isotopic and other markers suggesting a C3 plant diet. *Cretaceous Research* 24, 743–50.

Glennan, S. (2010). Ephemeral mechanisms and historical explanation. *Erkenntnis* 72(2), 251–66.

Godfrey-Smith, P. (2008) Recurrent, Transient Underdetermination and the Glass Half-Full. *Philosophical Studies* 137, 141–8.

Goodwin, M. B., Buchholtz, E. A., & Johnson, R. E. (1998). Cranial anatomy and diagnosis of Stygimoloch spinifer (Ornithischia: Pachycephalosauria) with comments on cranial display structures in agonistic behavior. *Journal of Vertebrate Paleontology* 18(2), 363–75.

Gould, S., & Eldredge, N. (1993). Punctuated equilibrium comes of age. *Nature* 366(6452), 223.

Gould, S. J. (1980). The promise of paleobiology as a nomothetic, evolutionary discipline. *Paleobiology* 6(1), 96–118.

Gould, S. J. (1965). Is uniformitarianism necessary? *American Journal of Science* 263(3), 223–8.

Green, R. E., Braun, E. L., Armstrong, J., Earl, D., Nguyen, N., Hickey, G., . . . & Kern, C. (2014). Three crocodilian genomes reveal ancestral patterns of evolution among archosaurs. *Science* 346(6215), 1254449.

Grimaldi D. (1999). The co-radiations of pollinating insects and angiosperms in the Cretaceous. *Annals of the Missouri Botanical Garden* 86, 373–406.

Guala, F. (2002). Models, Simulations, and Experiments. In L. Magnani & N. J. Nersessian (eds). *Model-based Reasoning: Science, Technology, Values.* New York: Kluwer Academic Publishers, pp. 59–74.

Hacking, I. (1983). *Representing and intervening.* (Vol. 279). Cambridge: Cambridge University Press.

Hawkes, C (1954). Archeological Theory and Method: Some Suggestions from the Old World. *American Anthropologist* 56, 155–68.

Hawley, K. & Bird, A. (2011). What are Natural Kinds? *Philosophical Perspectives* 25, 205–21.

Havstad, J. (2019). Let me tell you 'bout the birds and the bee-mimicking flies and Bambiraptor. *Biology & Philosophy* 34(25), 1-25.

Hedges, S. B., Parker, P. H., Sibley, C. G. & Kumar, S. (1996). Continental breakup and the ordinal diversification of birds and mammals.*Nature* 381:226–9.

Hempel, C. G. (1942). The function of general laws in history. *The Journal of Philosophy* 39(2), 35–48.

Horner, J. R., & Goodwin, M. B. (2006). Major cranial changes during Triceratops ontogeny. *Proceedings of the Royal Society of London B: Biological Sciences* 273(1602), 2757–61.

Hull, D. L. (1976). Are species really individuals? *Systematic Zoology* 25(2), 174–91.

Hull, D. L. (1975). Central subjects and historical narratives. *History and Theory* 14(3), 253–74.

Inkpen, R., & Turner, D. (2012). The topography of historical contingency. *Journal of the Philosophy of History* 6(1), 1–19.

Jackson, F., & Pettit, P. (1992). In defense of explanatory ecumenism. *Economics & Philosophy* 8(1), 1–21.

Jeffares, B. (2008). Testing times: regularities in the historical sciences. *Studies in History and Philosophy of Biological and Biomedical Sciences* 39(4), 469–75.

Jones, E. (2019). Ancient genetics to ancient genomics: celebrity and credibility in data-driven practice. *Biology & Philosophy* 34(27), 1-35.

Kehew, A. E., & Teller, J. T. (1994). History of late glacial runoff along the southwestern margin of the Laurentide ice sheet. *Quaternary Science Reviews* 13(9–10), 859–77.

Kosso, P. (2001). *Knowing the past: Philosophical issues of history and archaeology*. New York: Humanity Books.

Krause, J., Fu, Q., Good, J. M., Viola, B., Shunkov, M. V., Derevianko, A. P., & Pääbo, S. (2010). The complete mitochondrial DNA genome of an unknown hominin from southern Siberia. *Nature* 464(7290), 894.

Kuhn, T. S. (1970). The structure of scientific revolutions. Chicago: University of Chicago Press, pp. 84–5.

Laudan, L. (1990). Demystifying Underdetermination. In C. Wade Savage (ed.), *Scientific Theories, (Series: Minnesota Studies in the Philosophy of Science, vol. 14)*, Minneapolis: University of Minnesota Press, pp. 267–97.

Leakey, R and Lewin, R. (1992) *Origins Reconsidered: In Search of What Makes Us Human*. New York: Anchor.

Le Bihan, S. (2016). Enlightening Falsehoods: A Model View of Scientific Understanding. In S. R. Grimm, C. Baumberger & S. Ammon (eds), *Explaining Understanding: New Perspectives from Epistemology and Philosophy of Science*. Routledge, pp 111–35.

Leonelli, S. (forthcoming). The Time of Data: Time-Scales of Data Use in the Life Sciences. Philosophy of Science.

Leonelli, S. (2016). *Data-centric biology: a philosophical study*. Chicago: University of Chicago Press.

Lloyd, G. T., Davis, K. E., Pisani, D., Tarver, J. E., Ruta, M., Sakamoto, M., ... & Benton, M. J. (2008). Dinosaurs and the Cretaceous terrestrial revolution. *Proceedings of the Royal Society of London B: Biological Sciences* 275(1650), 2483–90.

Lyell, C. (1837). *Principles of Geology: Being an Inquiry How Far the Former Changes of The Earth's Surface are Referable to Causes Now in Operation* (Vol. 1). Philadelphia: J. Kay, Jun & Brother.

McConwell, A. (2019). Contingency's causality and structural diversity. *Biology & Philosophy* 34(26), 1-26.

McConwell, A. K., & Currie, A. (2017). Gouldian arguments and the sources of contingency. *Biology & Philosophy* 32(2), 243–61.

Machamer, P. K., Darden, L., & Craver, C. F. (2000). Thinking about Mechanisms. *Philosophy of Science* 67, 1–25.

Maclaurin, J., & Sterelny, K. (2008). *What is biodiversity?* Chicago: University of Chicago Press.

Marshall, C. R. (2017). Five palaeobiological laws needed to understand the evolution of the living biota. *Nature Ecology & Evolution*, 1(6), 0165.

Mäki, U. (2005), Models are Experiment, Experiments are Models. *Journal of Economic Methodology* 12(2), 303–15.

Meredith, R. W., Janecka, J. E., Gatesy, J., Ryder, O. A., Fisher, C. A., Teeling, E. C., … & Rabosky, D. L. (2011). Impacts of the Cretaceous Terrestrial Revolution and KPg extinction on mammal diversification. *Science* 1211028.

Millstein, R. L. (forthcoming). Types of Experiments and Causal Process Tracing: What Happened on the Kaibab Plateau in the 1920s. Studies in History and Philosophy of Science Part A.

Mitchell, S. (1997). Pragmatic laws. *Philosophy of Science* 64 (4), 479.

Mitchell, J. S., Roopnarine, P. D., & Angielczyk, K. D. (2012). Late Cretaceous restructuring of terrestrial communities facilitated the end-Cretaceous mass extinction in North America. *Proceedings of the National Academy of Sciences* 109(46), 18857–61.

Mink, L. O. (1978). Narrative form as a cognitive instrument. In L. Mink, R. Canary, & H. Kozicki (eds), *The writing of history: Literary form and historical understanding*. Madison: University of Wisconsin Press, pp. 129–49.

Morgan, M (2005). Experiments versus models: New phenomena, inference and surprise. *Journal of Economic Methodology* 12 (2), 317–29.

Nersessian, N. J. (2007). Thought experimenting as mental modeling: Empiricism without logic. *Croatian Journal of Philosophy* 7(20), 125–54.

Nersessian, N. (1999) Model-based reasoning in conceptual change. In L. Magani, N. Nersessian, & P. Thagard (eds), *Model-based reasoning in scientific discovery*. New York: Kluwer/Plenum, pp. 5–22.

Odenbaugh, J. (2006). Message in the bottle: The constraints of experimentation on model building. *Philosophy of Science* 73(5), 720–9.

Oppenheim, P & Putnam, H. (1958). Unity of Science as a Working Hypothesis. In H. Feigl, M. Scriven, & G. Maxwell (eds),*Concepts, Theories, and the Mind-Body Problem. Minnesota Studies in the Philosophy of Science, Volume II*. Minneapolis: University of Minnesota Press, pp. 3–36

Oreskes, N. (1999). *The rejection of continental drift: Theory and method in American earth science*. Oxford: Oxford University Press.

Paleobiology Research Group (accessed 23/11/2018). The Cretaceous Terrestrial Revolution. http://palaeo.gly.bris.ac.uk/macro/supertree/KTR.html

Parke, E. (2014) Experiments, Simulations, and Epistemic Privilege. *Philosophy of Science* 81 (4), 516–36.

Peterson, J. E., & Vittore, C. P. (2012). Cranial pathologies in a specimen of Pachycephalosaurus. *PloS One* 7(4), e36227.

Plutynski, A. (2018). Speciation Post Synthesis: 1960–2000. *Journal of the History of Biology*, 1–28.

Polanyi, M. (1958). *Personal knowledge*. Routledge.

Potochnik, A. (2017). *Idealization and the Aims of Science*. Chicago: University of Chicago Press.

Powell, R., & Mariscal, C. (2014). There is grandeur in this view of life: the bio-philosophical implications of convergent evolution. *Acta Biotheoretica* 62, 115

Ricoeur, P. (2010). *Time and narrative*. Vol. 3. Chicago: University of Chicago Press.

Roth, P. A. (2017). Essentially narrative explanations. *Studies in History and Philosophy of Science Part A* 62, 42–50.

Roth, P. A. (1988). Narrative explanations: the case of history. *History and Theory*, 1–13.

Rudwick, M. (1972). *The Meaning of Fossils: Essays in the History of Paleontology*. Chicago: University of Chicago Press.

Rudwick, M. J. (2014). *Earth's Deep History: How it was Discovered and why it Matters*. Chicago: University of Chicago Press.

Russell, B. (1921). *The Analysis of Mind*. Duke University Press.

Scannella, J. B., Fowler, D. W., Goodwin, M. B., & Horner, J. R. (2014). Evolutionary trends in Triceratops from the Hell Creek Formation, Montana. *Proceedings of the National Academy of Sciences*, 201313334.

Scannella, J. B., & Horner, J. R. (2010). Torosaurus Marsh, 1891, is Triceratops Marsh, 1889 (Ceratopsidae: Chasmosaurinae): synonymy through ontogeny. *Journal of Vertebrate Paleontology* 30(4), 1157–68.

Schott, R. K., Evans, D. C., Goodwin, M. B., Horner, J. R., Brown, C. M., & Longrich, N. R. (2011). Cranial ontogeny in Stegoceras validum (Dinosauria: Pachycephalosauria): a quantitative model of pachycephalosaur dome growth and variation. *PLoS One* 6(6), e21092.

Sterelny, K. (2016). Contingency and history. *Philosophy of Science* 83(4), 521–39.

Sterelny, K. (1996). Explanatory pluralism in evolutionary biology. *Biology and Philosophy* 11(2), 193–214.

Sullivan, R. M. (2006). The shape of Mesozoic dinosaur richness: a reassessment. *New Mexico Museum of Natural History and Science Bulletin* 35, 403–5.

Sullivan, R. M. (2003). Revision of the dinosaur Stegoceras lambe (Ornithischia, Pachycephalosauridae). *Journal of Vertebrate Paleontology* 23(1), 181–207.

Tucker, A. (2011). Historical science, over-and underdetermined: A study of Darwin's inference of origins. *The British Journal for the Philosophy of Science* 62(4), 805–29.

Turner, D. (forthcoming). In defence of living fossils. Biology & Philosophy.

Turner, D. (2017). Paleobiology's uneasy relationship with the Darwinian tradition: stasis as data. In R. G. Delisle (ed.), *The Darwinian Tradition in Context*. Basel: Springer, pp. 333–52.

Turner, D. (2016). A second look at the colors of the dinosaurs. *Studies in History and Philosophy of Science Part A* 55, 60–8.

Turner, D. (2013). Historical geology: Methodology and metaphysics. *Geological Society of America Special Papers* 502(2), 11–18.

Turner, D. (2007). *Making prehistory: Historical science and the scientific realism debate*. Cambridge: Cambridge University Press.

Turner, D. (2005). Local underdetermination in historical science. *Philosophy of Science* 72(1), 209–30.

Walsh, K. (forthcoming). Newton's Scaffolding: the instrumental roles of his optical hypotheses. Vanzo, A and Anstey, P (eds.), Experiment, Speculation and Religion in Early Modern Philosophy, Routledge.

Weisberg, M. (2007). Three kinds of idealization. *The Journal of Philosophy* 104(12), 639–59.

White, H. V. (1966). The burden of history. *History and Theory* 5(2), 111–34.

Wigner, Eugene (1960). The Unreasonable Effectiveness of Mathematics in the Natural Sciences. *Communications On Pure and Applied Mathematics* vol XIII, 1–14.

Wimsatt, W. C. (2007). *Re-engineering philosophy for limited beings: Piecewise approximations to reality*. Boston, MA: Harvard University Press.

Wylie, A. (2011). Critical distance : stabilising evidential claims in archaeology. In P. Dawid, W. Twining & M. Vasilaki (eds), *Evidence, Inference and Enquiry*. OUP/British Academy.

Wylie, A. (1999). Rethinking unity as a "working hypothesis" for philosophy of science: How archaeologists exploit the disunities of science. *Perspectives on Science* 7(3), 293–317.

Wylie, A. (2017). How archaeological evidence bites back: strategies for putting old data to work in new ways. *Science, Technology, & Human Values*, 42(2), 203-25.

Wylie, C. D. (2015). 'The artist's piece is already in the stone': Constructing creativity in paleontology laboratories. *Social Studies of Science* 45(1), 31–55.

Wylie, C. D. (2019). Overcoming the underdetermination of specimens. *Biology & Philosophy* 34(24), 1-18.

Acknowledgements

This Element has benefited from kind and incisive readers: Aaron Rackley, Alison McConwell, Andrea Raimondi, Daniel Swaim, Flavia Fabris and Kirsten Walsh, as well as two anonymous referees and the folks at a workshop at UBC. I'd like to thank Robert Northcott and Jacob Stegenga for dedicated editing work. I'm greatly indebted to the students of the University of Exeter's 2018 module *Deep Past, History and Humanity*, who graciously suffered through a draft and played an important role in shaping it. Aspects have also been presented at the University of Exeter Philosophy Society and the London School of Economics, many thanks to the audiences there.

I dedicate this Element to Kirsten Walsh, who matters.

Cambridge Elements ☰

Philosophy of Science

Robert Northcott
Birkbeck, University of London

Robert Northcott is Reader in Philosophy. He began at Birkbeck in the summer of 2011, and in 2017 became Head of Department. Before that, he taught for six years at the University of Missouri-St Louis. He received his PhD from the London School of Economics. Before switching to philosophy, Robert did graduate work in economics, receiving an MSc, and undergraduate work in mathematics and history.

Jacob Stegenga
University of Cambridge

Jacob Stegenga is a Reader in the Department of History and Philosophy of Science at the University of Cambridge. He has published widely on fundamental topics in reasoning and rationality and philosophical problems in medicine and biology. Prior to joining Cambridge he taught in the United States and Canada, and he received his PhD from the University of California San Diego.

About the Series

This series of Elements in Philosophy of Science provides an extensive overview of the themes, topics and debates which constitute the philosophy of science. Distinguished specialists provide an up-to-date summary of the results of current research on their topics, as well as offering their own take on those topics and drawing original conclusions.

Cambridge Elements ≡

Philosophy of Science

Elements in the Series

A full series listing is available at: www.cambridge.org/EPSC